51 Topics in Current Chemistry

Fortschritte der chemischen Forschung

Silicon Chemistry II

Springer-Verlag
Berlin Heidelberg GmbH 1974

This series presents critical reviews of the present position and future trends in modern chemical research. It is addressed to all research and industrial chemists who wish to keep abreast of advances in their subject.

As a rule, contributions are specially commissioned. The editors and publishers will, however, always be pleased to receive suggestions and supplementary information. Papers are accepted for "Topics in Current Chemistry" in either German or English.

Any volume of the series may be purchased separately.

ISBN 978-3-662-15558-5 ISBN 978-3-540-38052-8 (eBook)
DOI 10.1007/978-3-540-38052-8

Originally published by Springer-Verlag Berlin Heidelberg New York in 1974
Softcover reprint of the hardcover 1st edition 1974

Library of Congress Catalog Card Number 51-5497.

Contents

Springer-Verlag	D-6900 Heidelberg 1 · Postfach 1780 Telephone (06221) 49101 · Telex 04-61723 D-1000 Berlin 33 · Heidelberger Platz 3 Telephone (030) 822011 · Telex 01-83319
Springer-Verlag New York Inc.	New York, NY 10010 · 175, Fifth Avenue Telephone 673-2660

Properties and Preparations of Si–Si Linkages

In Inorganic and Organic Silicon Compounds

Professor Dr. Edwin Hengge

Institut für Anorganische Chemie der Technischen Hochschule, Graz, Österreich

Contents

1. Introduction

This review is an attempt to summarize present knowledge of the Si–Si linkage. The most important properties of a compound are conferred by the chemical properties of the bonds present in the molecule, so that a study of chemical bonds is very rewarding. The Si–Si bond is especially interesting from both the theoretical and practical points of view. Silicon is one of the most abundant and versatile elements and its compounds range from the silicates one of the most stable groups of terrestrial compounds to the extremely air-sensitive silanes, from transparent quartz to purest silicon for transistors, to polysiloxane lubricants, building protection agents, etc. Pharmacological activity was recently observed in organosilicon compounds. Silicon being the second most abundant element in the earth's crust, no shortage of supply can be expected.

Do we now know enough of theoretical and practical interest regarding silicon compounds? The answer is, we do not. Many of the silicon-containing compounds which today are of great technological importance were unknown a few decades ago and have originated from pure research. This basic research has generated many unanswered questions so that there is still much to be done in this field.

The chemistry of silicon is quite different from that of carbon, though there are some similarities. This comparison is significant because the two elements are close together in the periodic table, but also because carbon is the essential element in organic compounds and silicon is ubiquitous in the inorganic structure of the earth. The C–C bond is the most important organic linkage and the Si–O–Si bond the most important inorganic one. What caused these different bond systems to develop? Let us look at the properties of the respective elements: the C–C bond is more stable than the Si–Si bond because the atom is smaller and additional double bonds can be formed. Double bonds involving p orbitals cannot be formed with silicon, yet the Si–O–Si bond is more stable than a C–O–C bond thanks to a $d_\pi - p_\pi$ effect.

A direct comparison of the C–C bond with the Si–Si bond requires a thorough knowledge of the properties of the Si–Si bond.

The most important results that throw light on the Si–Si bond are summarized in this review. It deals first with the hydrogen and halogen compounds having Si–Si bonds, describes the preparation and properties of organic substituted disilanes and oligosilanes, and discusses the physicochemical properties of the Si–Si bond.

The oxidation state of a silicon atom in a chain of silicon atoms is two. Thus the chemistry of silicon in the oxidation state two assumes importance. Recent studies of the chemistry of silylenes in connection with polysilanes have produced significant results and the controversial silicon monoxide is accorded its rightful place in the systematic chemistry of silicon.

This review would be incomplete without a summary of the properties of the Si–Si bond in cyclic and heterocyclic silanes and polymeric compounds containing Si–Si bonds. Important results were obtained, *e.g.* the reason for the color of these compounds was discovered. The bibliography is very comprehensive and will lead the reader to a more detailed study of the subject. The author makes no claim to have covered the literature completely.

2. Silicon Hydrides with Si–Si Bonds

The simplest imaginable compound with Si–Si bonding is the disilane Si_2H_6. This disilane was discovered by Moissan in 1902[420–422]. A purer preparation was made by Lebeau *et al.*[33, 395] but it was first studied in detail by Stock *et al.* The disilane was obtained in high purity from the reaction of Mg silicide with hydrochloric acid, and major physicochemical parameters were measured[598, 603, 683]. Stock also demonstrated the reaction with gaseous hydrogen chloride and bromide forming halogen-containing derivatives[599], and the extreme sensitivity of the disilane to oxygen[601] and aqueous alkali. All these properties are essentially determined by the Si–H bonding. Si_2H_6 decomposes at 400–500 °C into SiH_4 and polymers. This tendency to disproportionate splitting into monosilanes and polymer compounds is frequently found in silicon chemistry, and is indicative of the relatively low strength of the Si–Si bonding and the stability of polymeric structures.

Attempts to produce higher silicon hydrides from H_3SiCl or H_2SiCl_2 by means of a Wurtz synthesis (with Na/K alloy or amalgam) were unsuccessful, SiH_4 and yellow polymeric silicon hydrides being formed for the reasons mentioned above.

Besides Si_2H_6, Stock found a series of siliconhydrides with longer chains of the general formula Si_nH_{2n+2} ($n < 7$); higher silanes seemed to be present but could not be isolated[605]. The extraordinary difficulties experienced in investigating the higher silanes (they undergo explosive reactions with very small amounts of oxygen, decompose in sunlight, and react with grease and similar materials) caused further experiments in this direction to be delayed until better experimental techniques had been developed.

Preparation of the higher silicon hydrides by reaction of Mg silicide with aqueous acids gives only small yields of silane mixtures. Decomposition of Mg silicide in ammonia with NH_4Br was successfully attempted, giving yields of silanes three times greater than with decomposition in aqueous medium. The content of higher silanes in the product depends on the composition of the Mg silicide and on reaction temperature, which should be low as possible for preparing higher silanes[323]. When hydrazine is used as the reaction medium[148], monosilane is preferentially obtained from hydrazinium chloride and Mg silicide.

The preparation of silicon hydrides became much easier in 1947[149], when it was found that halides can be reduced with $LiAlH_4$ and that higher silicon halides can also be reduced to the corresponding hydrogen compounds:

$$Si_2Cl_6 \xrightarrow{LiAlH_4} Si_2H_6 \qquad ^{149)}$$

$$Si_3Cl_8 \xrightarrow{LiAlH_4} Si_3H_8 \qquad ^{163)}$$

Nevertheless, decomposition of higher silanes does occur in this reaction. Besides these advances in preparation methods, progress has also been made in isolating higher silanes which are difficult to separate, so that the possibility of the existence of higher silanes and isomers could be further explored.

Emeleus and Maddock[408)] could not find isomers in Si_4H_{10} isolated by fractional distillation of the raw silane mixture. Feher *et al.*[139)] first demonstrated the existence of n- and i-Si_4H_{10} and Si_5H_{12} by Raman spectroscopy. The first isolations in pure form were made by Borer and Phillips[48)], Feher and Strack[147)] and Gokhale and Jolly[231)] using gas chromatography. The methods of preparation and handling silanes developed by Feher *et al.*[136–138, 145, 146)] made larger quantities available, enabling the existence of linear and branched-chain silicon hydrides up to $Si_{15}H_{32}$ to be demonstrated by gas chromatography, nmr, and Raman spectroscopy[144)].

Methods other than decomposition of silicides or hydrogenation of halodisilanes available for the synthesis of higher silicon hydrides are silent electrical discharge[112, 230, 531)] and pyrolysis of SiH_4 and Si_2H_6 to produce mixtures of higher silanes[51, 389, 460, 470)]. Investigations of the thermal decomposition of higher silanes show the presence of SiH_4 in addition to polymers[51, 606, 627)]. Studies of reaction mechanism show that equilibria are adjusted by means of silyl radicals[125)], but that silylene addition reactions and decomposition are also involved[606)] (see also Chapter 7; Silylenes). Single Si atoms in the excited state can also be insered into an Si–H bond[520)], though insertion into an Si–Si bond seems to be preferred[80)].

All previous investigations had demonstrated the existence of branched and linear chained silanes, though cyclic compounds had not so far been isolated until the cyclic silane, Si_5H_{10}, was recently prepared by Hengge *et al.*[255)] by hydrogenation of Si_5Br_{10}, obtained by interaction of HBr with $Si_5\varphi_{10}$, a known compound. Si_5H_{10} is a colorless liquid and is extremely sensitive to air. The boiling point was determined by extrapolation as about 130 °C. ^{1}H-NMR and Raman spectra were measured (only one singlet at 6.57 ppm, coupling constant 195 c/s, ring breathing frequency 382 cm^{-1}, molecular weight by mass spectroscopy : 150).

The chemical properties of the silicon hydrides are of course determined by the character of the Si–Si and Si–H bonding.

The Si–Si bond is relatively weak, so homolytic decomposition readily occurs on application of thermal or electrical energy or radiation (light, neutrons, X-ray . . .)[a)]. As expected from the coordinative unsaturated character of the Si atoms, the polysilanes are very reactive with nucleophilic agents as shown

a) See methods of preparation.

for instance by the familiar reaction with OH^- ions. Si–Si bonds are also very sensitive to oxidation, because the Si–O–Si bonding is thermodynamically strongly favoured. Since the Si–H bonding is also slightly susceptible to oxidation, the silanes are very readily oxidized and are strong reducing agents. Higher silanes spontaneously react with air, igniting or exploding; the explosive conditions have been determined[126].

Studies of reaction mechanisms by means of pyrolysis, silent electrical discharge, or neutron rays reveated that the original radicals or silylenes will react with a series of gaseous substances to produce new silane derivatives. Some examples are:

silicon hydrides + {
- $C_2H_2 \longrightarrow$ carbosilanes (see the review of G. Fritz[156a])
- $NH_3 \longrightarrow$ formation of Si–N bonds [113]
- germanium $\longrightarrow$ formation of Si–Ge bonds [11, 65, 405]
- $PH_3 \longrightarrow Si_2H_5PH_2$, $(H_3Si)_2PH$ [104, 106, 109, 162, 164, 165, 232, 233]
- $AsH_3 \longrightarrow Si_2AsH_7$, etc. [109]
- $PH_3/B_2H_6 \longrightarrow H_3Si \cdot SiH_2 \cdot PH_2 \cdot BH_3$ [104]

Higher silicon hydrides with Ge, *e.g.* $Si_2H_5GeH_3$ and other compounds produced in these pyrolysis reactions, have also been obtained by hydrolysis of Mg or Ca germanide-silicides[446, 475, 636] and have been separated by gas chromatography[446, 447]. P and As derivatives produced by pyrolysis also be obtained in coupling reactions:

$$H_3SiMH_2 + Si_2H_5X \longrightarrow SiH_3X + Si_2H_5MH_2 \qquad M = P, As;\; X = Cl, Br$$ [107]

$$Si_2H_5Br + LiAl(AsH_2)_4 \longrightarrow Si_2H_5AsH_3$$ [6]

$$Si_2H_5Br + LiAl(PH_2)_4 \longrightarrow Si_2H_5PH_2.$$ [435]

The Si–H bond in SiH_4 reacts with potassium in 1,2 dimethoxyethane giving the SiH_3K found by Ring and Ritter[471] in 1960. This silylpotassium does afford a synthesis of disilane:

$$H_3SiH + BrSiH_3 \longrightarrow Si_2H_6 + KBr$$ [377]

although a series of secondary reactions takes place, as found by Ring *et al.* Metallic potassium[161], KH[469] and salts of alkali metals similarly decompose disilane (presumably by splitting off a silylene group) and SiH_4 and polysilanes are produced[304, 426]:

$$xSi_2H_6 \xrightarrow{\text{salt}} xSiH_4 + {}^1/x\,(SiH_2)_x.$$

The catalytic effect of the salts in this disproportionation decreases in the order KH > LiD > LiCl[304]. Ring *et al.*[468] decided that the reaction is a bimolecular process (S_{N1}–Si mechanism) in which the activated complex is:

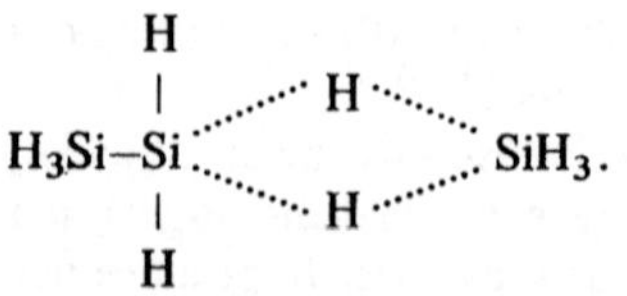

Silicon hydrides can be halogenated in different ways. As mentioned, Si_2H_6 reacts with HCl, HBr or HI in presence of aluminium halides to mono- and dihalodisilanes[1, 599, 662]. Hydrolysis of these monohalogen derivatives gives disilanylether $(Si_2H_5)_2O$, which can be split with BCl_3 into monochlorodisilanes[118]. The halogenation can be effected with free halogens (Cl_2, Br_2 [143], I_2 [141]), and cleavage of the Si–Si bonds can be avoided by the use of mild reaction conditions.

The homologous compounds $(Si_2H_5)_2S$ and $(Si_2H_5)_3N$ have been prepared and described in detail:

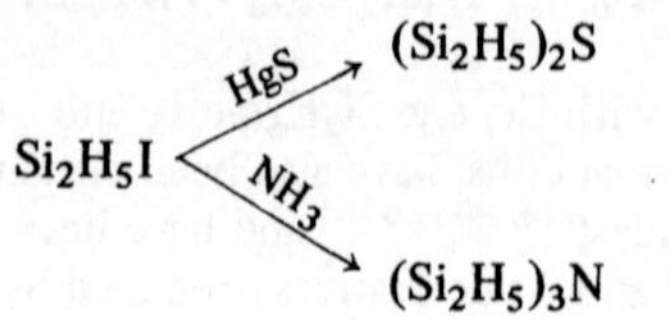

[664]

Ammonolysis of a mixture of Si_2H_5Cl and $Si_2H_4Cl_2$ was reported to give a mixture of $(Si_2H_5)_3N$ and $(Si_2H_4)NH$[160], but the compounds were not isolated and characterized.

Tri- and tetrasilanes have been partly halogenated in the same way. Bromination of di- and trisilanes has been most recently carried out without any solvent[168], so facilitating isolation. BCl_3 and BBr_3 proved to be good halogenating media[103, 105, 120], while neither BF_3 [120] nor $SnCl_4$ [32] is suitable. AgCl and AgBr can both halogenate Si_2H_6 and derivatives such as $meSi_2H_5$ [648] with precipitation of silver

$$-\overset{|}{\underset{|}{Si}}-H + 2\,AgCl \longrightarrow -\overset{|}{\underset{|}{Si}}-Cl + HCl + 2\,Ag$$

and in the case of the Si_2H_6 formation of di- to tetrachlorosilanes[295, 296].

Chlorination of the Si–H bond with weak chlorinating agents such as CCl_4 or $CHCl_3$ appears feasible if the Si atoms of the Si–H bond can be further bonded to at least two other Si atoms. The IR valence frequency of the Si–H bond is found to be mostly under 2100 cm^{-1}. This reaction procedure turned out to be effective for halogenation of labile compounds[65, 265, 350, 604].

Partly fluorinated disilanes have been prepared by electrical discharge in $FSiH_3$[114]. Partly chlorinated and brominated disilane and trisilane derivatives may be so produced and have been characterized by spectroscopy[108].

Si_2H_5Br originates by bromination of Si_2H_5I with AgBr[663] and base-catalyzed conversion leads to $Si_5H_4Br_2$ and Si_2H_6[402].

3. Silicon Halides with Si–Si Bonds

In the disilane series, all representatives of the hexahalodisilanes Si_2X_6 are well known but none of the mixed disilanes substituted with different halogen atoms.

The methods of preparation of the hexahalodisilanes have long been known and are well reviewed[228, 251].

Si_2F_6, a colorless gas (b. p.: −19 °C), is usually obtained by fluorination of Si_2Cl_6[572, 573] or Si_2Br_6[549] but cleavages of the Si–Si bond can occur. Si_2F_6 can be separated more easily with Si_2Br_6 as starting material as the result of the greater difference between the boiling points.

Si_2Cl_6, a colorless liquid (b. p.: 147 °C) is normally obtained from the silicides of Ca or Fe by treatment with chlorine[416, 461, 574]. The presence of Mn or Cu[254, 638] improves the yield even at the lowest possible reaction temperature. Si_2Cl_6 can also be obtained by reaction of $SiCl_4$ and Si[640], or by reaction of chlorine with metallic silicon during grinding at room temperature which affords a fresh and reactive surface[354]. Si_2Cl_6 can also be prepared by chlorination of Si_2I_6 with $HgCl_2$ [5, 153] by heating of $HSiCl_3$ [35], and by electrical discharge in $SiCl_4$[9, 10, 597].

Si_2Br_6 is produced in the reaction of $CaSi_2$ with gaseous bromine[570] but also by treatment of Si and HBr[35]. It is a white crystalline compound (m.p.: 109 °C), which shows an allomorphic transition at 95 °C.

Si_2I_6 is obtained by reaction of SiI_4 with silver:

$$2\,SiI_4 + Ag \xrightarrow{290-300^\circ} Si_2I_6 + 2\,AgI \qquad {}^{152,\,587)}$$

or even by reaction of iodine with silicon[396]. Si_2I_6 is whitish-yellow, undergoes sublimation in vacuum above 150 °C, and is the most heat-sensitive compound among the halodisilanes; it decomposes in SiI_4 and polymers $(SiI)_n$ above 250 °C.

As regards their chemical properties, halodisilanes are (as expected) to be very reactive thermally labile nucleophiles. Acid hydrolysis occurs to give polymeric compounds (see p. 92) retaining Si–Si bonds; alkaline hydrolysis yields silicic acid with disruption of the Si–Si bonding. Partial hydrolysis carried out with Si_2Cl_6 led to Si_4OCl_{10}, $Si_6O_2Cl_4$ and $Si_8O_3Cl_{18}$ (see p. 93).

Si_2F_6 decomposes during acid or neutral hydrolysis to hexafluosilicic acid, hydrogen, and silica:

$$Si_2F_6 + 4\,H_2O \longrightarrow 6\,HF + H_2Si_2O_4$$
$$6\,HF + H_2Si_2O_4 \longrightarrow H_2 + H_2SiF_6 + H_4SiO_4$$
$$H_4SiO_4 + 6\,HF \longrightarrow 4\,H_2O + H_2SiF_6.$$

The reaction with ammonia has been investigated in detail by Billy. Since they lead chiefly to polymeric compounds, these reactions are reported in the chapter on polymeric compounds containing Si–Si bonds (see p. 93).

A reaction of special interest occurs with triethylamine and has been described by Urry *et al.*[324–327, 436, 646, 647]. Triethylamine catalyses conversion of Si_2Cl_6 to higher silicon halides such as Si_5Cl_{12} and Si_6Cl_{14}, besides $SiCl_4$[672]. Urry believes that adduct formation with me_3N initiates the reaction:

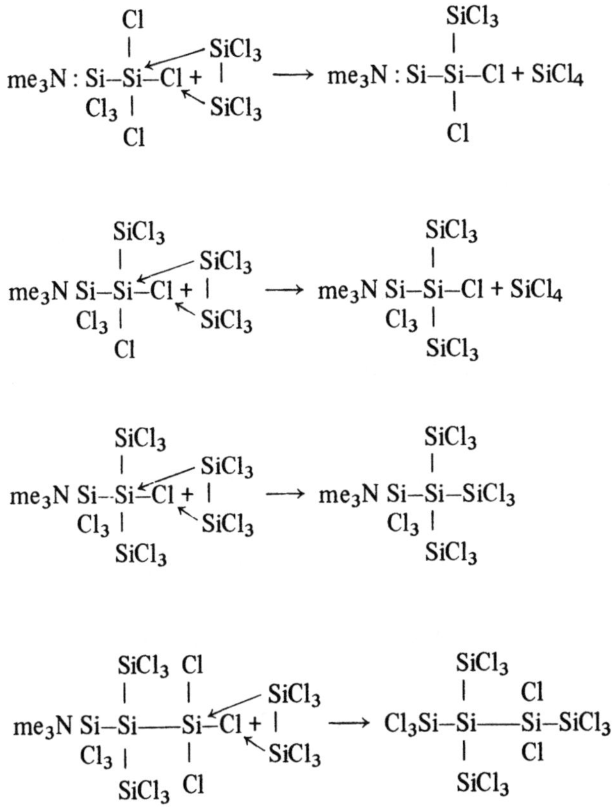

The reaction with Si_3Cl_8 similarly forms higher silicon halides of the following structures:

$$\begin{array}{c} SiCl_3 \\ | \\ Cl_3Si\text{–}Si\text{–}SiCl_3 \\ | \\ SiCl_3 \\ \hline Si_5Cl_{12} \end{array} \qquad\qquad \begin{array}{c} SiCl_3 \\ |\ \ Cl \\ Cl_3Si\text{–}Si\text{–}Si\text{–}SiCl_3 \\ |\ \ Cl \\ SiCl_3 \\ \hline Si_6Cl_{14} \end{array}$$

Corresponding disproportionation reactions have been found by Emeleus *et al.*[127] using ph_3As und ph_3P:

$$n\ Si_2Cl_6 \longrightarrow n\ SiCl_4 + (SiCl_2)_n.$$

Cleavages of the Si–Si bonding occur not only with fluorinating agents (see Si_2F_6, p. 10) but also with other halogens to yield mixed halogen-substituted monosilanes:

$$Si_2Cl_6 + Br_2 \xrightarrow{500\ °C} SiCl_3Br + SiCl_2Br_2.$$

The method of preparation of asymmetrical substituted disilanes (see p. 20) devised by Bettler, Sendra and Urry[645] is in principle suitable for preparing higher silicon halides. Thus it is possible to synthesize a pentachlorodisilanylmercury

$$2\ HSi_2Cl_5 + HgR_2 \longrightarrow 2\ RH + Hg(Si_2Cl_5)_2$$

which gives n-Si_4Cl_{10} quantitatively by photolysis:

$$Hg(Si_2Cl_5)_2 \xrightarrow{h\nu} Hg + Si_4Cl_{10}.$$

The starting material is not readily available but can be made by the reaction:

$$H_2SiCl_2 + Hg(SiCl_3)_2 \longrightarrow Hg + HSiCl_3 + HSi_2Cl_5.$$

This reaction again affords another opportunity for preparing higher silicon chlorides, for example:

$$H_2SiCl_2 + 2\ Hg(SiCl_3)_2 \longrightarrow Hg + 2\ HSi_2Cl_5 + \text{n-}Si_5Cl_{12}.$$

Branched chloropolysilanes can be obtained as follows:

$$H_3SiCl + 3\ Hg(SiCl_3)_2 \longrightarrow Hg + 3\ HSiCl_3 + \begin{array}{c} SiCl_3 \\ | \\ Cl_3Si\text{–}Si\text{–}SiCl_3. \\ | \\ Cl \end{array}$$

Higher silicon halides Si_nX_{n+2} with $n > 2$ but of limited molecular weight have been known for a long time, above all from chlorine, but a few from bromine, and none from iodine. In the case of fluorine, some higher silicon fluorides derived from difluorosilylene have been recently discovered (see p. 60).

Gattermann[166,167], Besson[34] and Martin[415] produced perchlorinated silicon chains with $n = 3$, 4 and 5, but the separation of the individual species by distillation is very difficult. Nowadays gas chromatographic methods enable separation[651]. Using the method of Schwarz *et al.*[580], $SiCl_4$ is converted in a hot-cold tube into a series of higher silicon halides (*e.g.* $Si_{25}Cl_{52}$) and from these it should be possible to obtain cyclic derivatives such as the bicyclic $Si_{10}Cl_{28}$[584–586].

Admittedly, their structures have not been investigated. In a hydrogen atmosphere, a halide of the composition $Si_{10}Cl_{20}H_2$ is formed[590].

These studies by Schwarz have been questioned by Kautsky and Kautsky[337, 344], who produced higher siliconhalides by an arc reaction in a mixture of Si and $SiCl_4$ suspension and found Si_6Cl_{14} as the highest distillable chloride, while all higher halides were thermally decomposed on distillation. Bonitz[47] obtained higher silicon halides by the reaction of Si with chlorine in a special fluidized-bed reactor, though they were not preasely characterized.

In summary, a series of higher (and presumably also cyclic) siliconhalides undoubtedly exists, but their actual structures seem uncertain at present.

Recently, a decabromocyclopentasilane Si_5Br_{10} was prepared in a completely different way (see p. 6) and its cyclic structure seems to have been proved by IR and Raman spectral investigations[255].

4. Organic Substituted Disilanes

Detailed investigations of organic derivatives of disilane have begun, particularly in the last years. Reviews by Kumada *et al.*[368] of aliphatic substituted disilanes have appeared, while older results are summarized in Gmelins' handbook[228]. Some details of these compounds are summarized by Gilman *et al.*[180, 190, 684], Schott[557] and MacKey and Watt[406].

The possibilities of synthesis are limited by the general properties of the Si–Si bond to a small number of reaction types. Disilane derivatives were usually synthesized by a Wurtz-like reaction of monosilane halide derivatives with alkali or alkaline-earth elements. Some pyrolysis reactions that have yielded disilane derivatives are also known. Often, new disilane derivatives were prepared from common derivatives by an exchange of substituents. Some rearrangements have been observed in recent years.

4.1. Preparations by Formation of the Si–Si Bond

The first organic substituted disilane was obtained by Friedel and Ladenburg in 1869 with a reaction of Si_2I_6 and $Znme_2$[154, 155]. Hexaalkyldisilanes are also formed in 50% yield by reaction of Si_2Cl_6 with Grignard reagent (R = et, pr, tolyl[577]). The preparation of hexamethyldisilane with this reaction is possible only in small yield because a cleavage reaction of the Si–Si bond occurs[69]. With $Znme_2$ this cleavage does not take place and the yield of Si_2me_6 increases to 85%[25, 155]. Besides the cleavage reaction, metal exchange and other coupling reactions occur. In the reaction of $SiCl_4$ with the cyclohexyl-Grignard compound for example, not only the expected cyclohexylmonosilane, but also $[(C_6H_{11})Cl_2Si]_2$ was found (Kipping and Palmer[441]).

These difficulties and generally low yields in the preparation of organic disilane derivatives by the reaction of disilane hexahalide with organometallic compounds lead to the conclusion that the better method for preparation of these compounds is the Wurtz-type synthesis starting from organohalomonosilane.

First preparations were made by Schlenk *et al.*[546] who synthesized $Si_2\varphi_6$ by the reaction of $Si\varphi_3Cl$ with sodium. Kipping[381] prepared the first mixed-substituent disilanes by reaction of the corresponding organochlorosilanes with sodium, but was not able to separate the optical isomers. Kraus and Nelson[391] obtained Si_2 et_6 by this method.

However, the limitations of these methods were soon discovered, and Kipping[385] found that the reaction of $\varphi_2(O\varphi)$ SiCl with Na does not yield the expected disilane derivative. A Si–O–C cleavage occurs and some reaction products $(Si\varphi_2)_4$, $O\varphi\,(Si\varphi_2)\,O\varphi$ and $\varphi_2\,(O\varphi)_2Si$ result, perhaps via radicals. Similar observations were sometimes made on the cleavage reaction of the Si–O–C bond systems:

$$\varphi_3Si(Ome) \xrightarrow[\varphi_3SiCl]{Na/K} Si_2\varphi_6 + \text{polymers} \qquad {}^{30)}$$

$$\varphi_3Si(Oet) \xrightarrow{Na} Si\varphi_4 + \text{polymers} \qquad {}^{226)}$$

$$me_2Si(Ome)_2 \xrightarrow{Na/K} Si(Ome)_4 + \text{polymers.} \qquad {}^{273)}$$

Formation of $(\varphi O)_6Si_2$ alone was found in the reaction of sodium with $(O\varphi)SiBr$ and $(O\varphi)_2SiBr$, respectively[448]. This result contrasts with the results of Kipping. He found only derivatives of monosilane and metallic silicon in an analogous reaction with the chlorides.

Metal-exchange reactions were also observed, as in the reaction of phenyl-chlorosilanes and phenyl-silylhydrides, when products in addition to the expected $Si\varphi_4$ were formed[195]. This may be due to the formation of metal-silyl compounds during the reaction. It is known that organometallic compounds are easily formed with phenyl-substituted silane derivatives and with the Si–H bond.

Wurtz-like reactions were carried out with many organohalosilanes and it is not possible in this review to present details on all the compounds formed. Some examples of the general reaction

$$2\,R^1R^2R^3SiX \xrightarrow[\text{(Na/K or Na/Hg or Mg)}]{\text{alkali metal}} (R^1R^2R^3Si)_2$$

are summarized below:

$R^1 = R^2 = R^3 = R$	Refs.
$R = \varphi$	29, 173, 189, 375)
R = p-tolyl	57, 189)
p-biphenyl	189)
me	58, 214, 355, 535, 650)
et	391, 497, 650)
bn	650)
i-Am	650)
$-C_6H_4X$ ($X = F, Cl, CH_3, OCH_3$)	375)
C_6H_5	278)

R^1, R^2, R^3	Refs.
$R^1 = R^2 = me, R^3 = \varphi$	214)
$R^1 = R^2 = \varphi, R^3 = H$	
$R^1 = \varphi, R^2 = me, R^3 = H$	594)
$R^1 = \varphi, R^2 = et, R^3 = H$	
$R^1 = R^2 = me, R^3 = H, me, Ome, Nme_2\ [N\ me_2H]^+$	375)
$R^1 = R^2 = \varphi, R^3 = H, Cl, Br$, p-tolyl	680)
$R^1 = me, R^2 = R^3 = H$	95)
$R^1 = et, R^2 = pr, R^3 = Cl$	497)
$R^1 = \varphi, R^2 = R^3 = et$	86)

More complex organic groups do not interfere with the Wurtz reaction so long as no steric hindrance occurs.

The following reaction takes place:

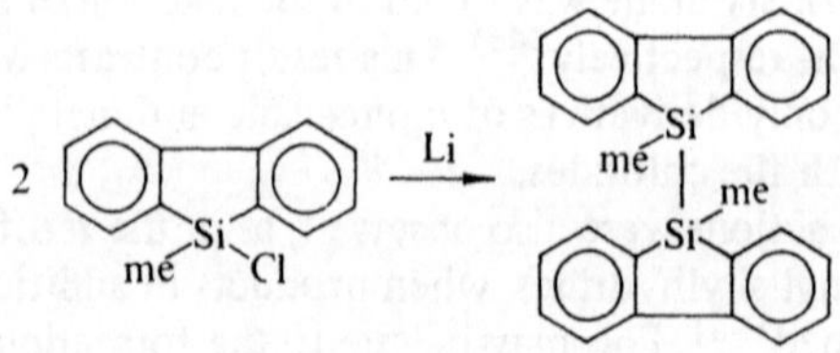

192)

but tris(cyclohexyl)chlorosilane does not react[431].

The results of the Wurtz-like reaction depend on the reaction conditions. An important factor is the kind of solvent. The ethers are generally better than hydrocarbons, but with the cyclic ethers like tetrahydrofuran (THF), dioxane and other frequently used agents, a cleavage reaction of the ethers was observed[392]. Hexamethylphosphoricacid amide was recently used with good results in combination with lithium[497].

Synthesis of asymmetric derivatives of disilanes is also possible:

$$\varphi_3SiCl + Mg + me_3SiCl \longrightarrow \varphi_3SiSime_3 \quad {}^{116)}$$

$$\varphi_2meSiCl + Li + ClSi\varphi_3 \longrightarrow \varphi_2meSiSi\varphi_3 \quad {}^{208)}$$

However, side reactions occur. The symmetric compounds were sometimes formed, probably due to formation of metal silicon compounds during the reaction.

The formation of disilane derivatives is not possible only with metals; organometallic compounds also react with halosilanes. West *et al.*[509] found formation of disilanes in the reaction with Grignard compounds:

$$\varphi_3SiCl \xrightarrow{RMgX} \varphi_6Si_2 .$$

West postulated a silyl magnesium compound (a silyl Grignard!) as the following reaction cannot be explained in any other way:

$$\varphi_3SiCl + me_3SiCl \xrightarrow{\text{hexyl–MgCl}} \varphi_3SiSime_3 .$$

In a similar manner, various phenylated disilanes were obtained by the reaction of me_3SiCl with C_6H_5MgCl in the presence of hexamethyldisilane[70].

Organosilicohydrides with one Si–φ bond or one trimethylsilyl group react with trimethylchlorosilane in the presence of secondary or tertiary Grignard compounds to form asymmetrically substituted disilanes:

$$R^1_2R^2SiH + ClSime_3 \xrightarrow{R^3MgCl} R^1_2R^2SiSime_3$$

$R^1 = \varphi \quad R^2 = H, me, \varphi,$ [299]

$R^1 = me \quad R^2 = Sime_3, \quad R^3 = -Cme_3, -CHme_2 .$

The Grignard compounds do not react directly with the starting materials, and the results can only be explained by the formation of a silyl magnesium compound during the reaction.

The mechanism of the Wurtz-like reaction in organosilicon chemistry is controversial. In some cases, for instance with phenylated derivatives, the formation of an alkali silicon compound may be significant, but the formation of a silylene during the reaction is generally assumed (see also Chapter 7: Silylenes). This intermediate occurs in the synthesis of pentamethyldisilane by Skell and Goldstein[518]. The carbon analog dimethylsilylene is formed as a short-lived compound carried in a helium gas stream after contact of dimethylchlorosilane gas in He with sodium/potassium alloy at 260–280°. This dimethylsilylene reacts as a dilute gas with trimethylsilane to give Si_2me_5H.

Disilanyl radicals are also observed. Pentamethyldisilane reacts with d-t-butylperoxide in toluene to form pentamethyldisilanyl derivatives such as (o, m, p-)meC_6H_4–Si_2me_5, me_3CO–Si_2me_5 and others. The intermediate can only be the pentamethyldisilanyl radical[482]. However, formation of asymmetrically substituted disilanes with a Wurtz-like synthesis starting from a mixture of the halosilanes is not the best method for preparing these compounds. A better way is reaction with organoalkali compounds:

$$R^1_3SiM + ClSiR_3 \longrightarrow R_3SiSiR_3 \qquad M = Li, Na, K, etc.$$

but this method is generally limited by the need for a suitable organometallosilicon compound.

The first preparation of such an alkali-organosilicon compound was made by Kraus[391] and Benkeser[31]. Gilman and Wu[224] first cleaved hexaphenyldisilane with potassium/sodium alloy and obtained triphenylsilyl potassium as a dark solution in ether.

The preparation of the pure compound was not possible. However, in solution, asymmetric disilanes are formed if triphenylsilyl potassium is treated with organosilicon halides. Some examples of these reactions are given in the following equations:

$$\varphi_3SiK + ClSime_3 \longrightarrow \varphi_3SiSime_3$$
$$\varphi_3SiK + ClSi\,et_3 \longrightarrow \varphi_3SiSi\,et_3 \quad \text{31, 224)}$$
$$\varphi_3SiK + (C_7H_7)_{3-n}\varphi_nSiCl \longrightarrow \varphi_3SiSi\varphi_n(C_7H_7)_{3-n} \quad \text{225)}$$
$$\varphi_3SiK + ClSi\varphi Cl_2 \longrightarrow \varphi_3SiSi\varphi Cl_2 \quad \text{225)}$$
$$\varphi_3SiK + ClSi\,et_2\varphi \longrightarrow \varphi_3SiSi\,et_2\varphi \quad \text{214)}$$
$$\varphi_3SiK + ClSime_n\varphi_{3-n} \longrightarrow \varphi_3SiSi\varphi_{3-n}me_n \quad \text{187, 202, 208)}$$
$$\varphi_3SiK + ClSi\varphi_2Cl \longrightarrow \varphi_3SiSi\varphi_2Cl \quad \text{227)}$$
$$\varphi_3SiK + Cl_4Si \longrightarrow \varphi_3SiSiCl_3 \quad \text{225)}$$
$$H_3SiK + ClSime_3 \longrightarrow H_3SiSime_3 \quad \text{14)}$$
$$\varphi_3SiK + (benzyl)\varphi_2SiCl \longrightarrow \varphi_3SiSi\varphi_2(benzyl). \quad \text{55)}$$

The expected products are not always obtained, *e.g.* when a metal rearrangement takes place and the reaction becomes more complicated, as in the reactions described by Gilman and coworkers[172]:

$$me_2\varphi SiLi + Si_2\varphi_6 \longrightarrow me_2\varphi SiSi\varphi me_2 + \varphi_3SiLi$$
$$me_2\varphi SiLi + me_2\varphi SiSi\varphi_3 \longrightarrow me_2\varphi SiSi\varphi me_2 + \varphi_3SiLi$$
$$me\varphi_2SiLi + me\varphi_2SiSi\varphi_3 \longrightarrow me\varphi_2SiSi\varphi_2me + \varphi_3SiLi$$

etc.

Gilman's significant observation that triphenylchlorosilane is a catalyst in Wurtz-like reactions forming cyclic silanes appears to be due to intermediate formation of triphenylsilyl lithium (see Chapter 9: Cyclic Silanes). The triphenyl compounds are particularly good catalysts because the corresponding alkali-triphenylsilyl compounds are the most stable among the alkali-organosilicon compounds.

In recently reported investigations by Kumada *et al.*[361, 366] a mechanism is postulated for the reaction with triphenylsilyl lithium as catalyst. Starting with α, α, α, ω, ω, ω-hexaphenylpolymethylpolysilane, the reaction yields polysilyl lithium compounds and may be represented by the equation:

$$\varphi_3Si(Sime_2)_nSi\varphi_3 + Li(Sime_2)_ZSi\varphi_3 \longrightarrow \sum_{x=0}^{x=6} \varphi_3Si(Sime_2)_XSi\varphi_3 + \sum_{y=0}^{y=6} Li(Sime_2)_YSi\varphi_3. \quad z = 0-4$$

If y is five or six, a cyclization occurs:

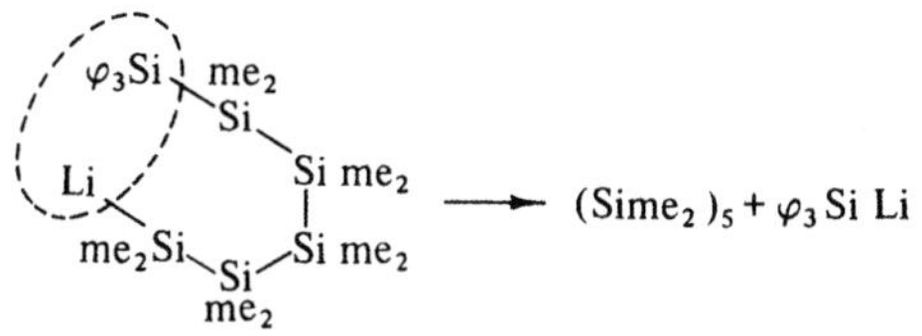

Complications in these reactions between an alkali-organosilico compound and a halogenated silane result from a frequent symmetrization of the disilanes formed, caused by a metal exchange. Kelling[375] found the following unexpected products:

$$\varphi_3SiK + (p\text{-}ClC_6H_4)_3SiCl \longrightarrow Si_2\varphi_6 + (p\text{-}ClC_6H_4)_6Si_2 .$$

If formation of silicon-element bonds by a reaction of the alkali-organosilicon compound and a halide of the corresponding element is attempted a disilane is usually formed:

$$\varphi_3SiK + \left\{\begin{matrix} AsCl_3 \\ SbCl_3 \\ BiCl_3 \end{matrix}\right. \longrightarrow \varphi_6Si_2 + \ldots \qquad {}^{170)}$$

$$\varphi_3SiLi + HgCl_2 \longrightarrow \varphi_6Si_2 + \ldots \qquad {}^{171)}$$

$$\varphi_3SiK + ClC\varphi_3 \not\longrightarrow \varphi_3SiC\varphi_3 \longleftarrow \varphi_3SiCl + NaC\varphi_3 \qquad {}^{56)}$$
$$\hookrightarrow Si_2\varphi_6 + C_2\varphi_6$$

Hexaphenyldisilane is also obtained in the reaction of triphenylsilane or triphenylmethoxysilane with Na/K alloy after the decomposition of triphenylsilyl lithium:

$$\varphi_3SiH \xrightarrow[\text{2.) } CH_3I]{\text{1.) Na/K}} \varphi_6Si_2 + \varphi_3Sime \qquad {}^{30)}$$

$$\varphi_3Si(Ome) \xrightarrow[\text{2.) } et_3SiCl]{\text{1.) Na/K}} \varphi_3SiSiet_3 . \qquad {}^{30)}$$

Polyhalogen compounds of methane react with triphenylsilyl lithium to form hexaphenyldisilane and the expected compounds do not occur[175–177].

It has been confirmed that the availability of such reactions depends upon the existence of the metal-organosilicon compound. The cleavage of disilanes is only possible if there is at least one phenyl group in the compound; *e.g.* it is not possible to cleave hexamethyldisilane.

The synthesis of trimethyl-[259] or triethylsilyl lithium[538, 653] may be effected via the corresponding mercury compound:

$$Hg(Sime_3)_2 \xrightarrow{Li} 2\ me_3SiLi + Hg$$
$$me_3SiLi + ClSi(Ome)_3 \longrightarrow me_3SiSi(Ome)_3. \quad ^{259)}$$

The easy decomposition of the mercury compounds can be used in direct synthesis of disilanes:

$$(R_3Si)_3Hg \xrightarrow{h\nu} (R_3Si)_2 + Hg.$$

A reaction mechanism via silyl radicals is plausible[121]. Bettler, Sendra and Urry[37, 645] treated bis(trimethylsilyl)mercury with $SiHCl_3$:

$$HSiCl_3 + Hg(Sime_3)_2 \longrightarrow Hg + HSime_3 + me_3SiSiCl_3.$$

A simple method was thus shown for the formation of 1,1,1-trimethyl-2,2,2-trichlorodisilane, previously formed only with great difficulty; the method might permit formation of similar compounds, for example, the series of compounds $Si_2me_{6-n}Cl_n$:

$$meSiHCl_2 + Hg(Sime_3)_2 \longrightarrow Hg + me_3SiH + me_3SiSimeCl_2$$
$$(me_3SiCH_2)_2Hg + 2\ me_2SiHCl_2 \xrightarrow{-2\ Sime_4} Hg(Sime_2Cl)_2 \longrightarrow Hg + $$
$$+ me_2ClSiSiClme_2$$

etc.

Formation of inorganic silicohalides via silicohalide-mercury compounds is also possible (see Chapter 3: Silicon Halides with Si–Si Bonds):

$$2\ H\ SiCl_3 + Hg[CH_2Sime_3]_2 \longrightarrow 2\ Sime_4 + Hg(SiCl_3)_2$$
$$2\ H\ Si_2Cl_5 + HgR_2 \longrightarrow 2\ RH + Hg(Si_2Cl_5)_2.$$

Diphenylmercury reacts with iododisilanes to form the corresponding phenyl derivatives $Si_2H_5\varphi$ and $(H_3Si)_2SiH\varphi$, probably via a disilanylmercury compound[142]. Other known disilanylmercury compounds[357, 652] are, for example:

$$et_3SiSiet_2H + et_2Hg \xrightarrow[19\ h]{160\ °C} C_2H_6 + etHgSiet_2Siet_3. \quad ^{652)}$$

A recent method for formation of alkali organosilicon compounds by cleavage of hexamethyldisilane with sodium methylate is described by Sakurai *et al.*[498]:

$$Si_2me_6 + NaOme \longrightarrow me_3SiNa + me_3SiOme.$$

The cleavage takes place only in hexamethylphosphoric acid triamide (HMPT) but this solvent hinders the reaction to the expected disilane with chlorosilanes. The chlorosilane reacts with the oxygen in the solvent, first forming a complex and then the corresponding disiloxane. Substituted phosphines are by-products[262]. A similar attempt to cleave n-propylpentamethylsilane leads to a rearrangement[496]:

$$2(\text{n-pr})\text{me}_2\text{SiSime}_3 \xrightarrow{\text{NaOme, HMPT}} [(\text{n-pr})\text{me}_2\text{Si}]_2 + (\text{me}_3\text{Si})_2 .$$

A practicable reaction mechanism is the nucleophilic attack by the silyl anion on the Si–Si bond, which leads to the symmetrical disilanes:

$$\text{me}_3\text{Si}^- + (\text{n-pr})\text{me}_2\text{SiSime}_3 \longrightarrow \text{Si}_2\text{me}_6 + (\text{n-pr})\text{me}_2\text{Si}^-$$
$$(\text{n-pr})\text{me}_2\text{Si}^- + (\text{n-pr})\text{me}_2\text{SiSime}_3 \longrightarrow \text{Si}_2(\text{n-prme}_2)_2 + \text{Sime}_3^- .$$

Lastly, a thermal Si/Si redistribution was observed by Sakurai and Hosomi[483]:

$$2\ \text{Rme}_2\text{SiSime}_3 \rightleftharpoons \text{me}_6\text{Si}_2 + (\text{Rme}_2\text{Si})_2$$

The symmetrical products are formed by a radical mechanism.

4.2. Preparation of Disilane Derivatives by Exchange of Substituents (Including Side Reactions)

A good general method for preparing organic derivatives of disilanes is the exchange of substituents, starting from simple derivatives with reactive substituents. One can expect difficulties with this method if the reagents are liable to cleave the Si–Si bond. Many side reactions, cleavages, rearrangements and other reactions which illustrate the reactivity of organically substituted disilanes are observed.

4.2.1. Reactions with Organometallic Compounds

4.2.1.1. Reactions with Grignard compounds

An obvious reaction for formation of organic derivatives of disilanes is the reaction of Grignard compounds with halogenated disilanes. But the Grignard reagent usually cleaves the disilane and only monosilane derivatives are observed[124, 203, 272, 413].

Generally, the Si–Si bond tends to be cleaved[261] when

a) the Si–Si bond is asymmetrically substituted by different substituents and the Si–Si bond becomes slightly polar.
b) more than two silicon atoms are grouped together, as in trisilanes, etc.,
c) in cyclosilanes which have a sufficiently large ring strain.

Under suitable reaction conditions, the Grignard reaction is a useful means of forming disilane derivatives. Some examples following the general reaction:

$$-\overset{|}{\underset{|}{Si}}-\overset{|}{\underset{|}{Si}}-X + RMgX \longrightarrow -\overset{|}{\underset{|}{Si}}-\overset{|}{\underset{|}{Si}}-R + MgX_2$$

should demonstrate that:

$$\left.\begin{matrix} Si_2Cl_6 \\ \\ Si_2Br_6 \end{matrix}\right. + \left\{\begin{matrix} C_3H_7MgBr \\ C_2H_5MgBr \\ \varphi MgBr \\ \text{tolyl MgBr} \end{matrix}\right\} \longrightarrow \text{corresponding organodisilanes } R_6Si_2 \qquad ^{577)}$$

$$(etO)_2meSiSime(Oet)_2 + meMgBr \longrightarrow \begin{matrix} me_3SiSime_2(Oet) \\ (Oet)me_2SiSime_2(Oet) \end{matrix} \qquad ^{371)}$$

$$Si_2Cl_6 + etMgBr \longrightarrow \begin{matrix} Si_2Cl_5et \\ Si_2Cl_4et_2 \end{matrix} \qquad ^{678)}$$

$$me_5Si_2Cl + viMgCl \longrightarrow viSi_2me_5 \qquad ^{499)}$$

$$Si_2Cl_6 + meMgBr \longrightarrow Si_2me_6 \qquad ^{69,\,414)}$$

$$Clme_2SiSime_2Cl + \varphi MgCl \longrightarrow \varphi me_2SiSime_2\varphi \qquad ^{358)}$$

$$\begin{matrix} Clme_2SiSime_3 \\ \\ Clme_2SiSime_2Cl \end{matrix} \xrightarrow{CH_2=\overset{\overset{me}{|}}{C}-MgBr} \begin{matrix} [CH_2=C(me)]me_2SiSime_3 \\ \\ [CH_2=C(me)]me_2SiSime_2[C(me)=CH_2] \end{matrix} \qquad ^{364)}$$

$$(Cl_2CH)me_n\,Cl_{5-n}Si_2 \xrightarrow{meMgBr} (Cl_2CH)me_nSi_2me_{5-n} \qquad ^{356)}$$

$$(etO)me_2SiSime_2(Oet) \xrightarrow{\varphi MgBr} \varphi me_2SiSime_2(Oet) \qquad ^{358)}$$

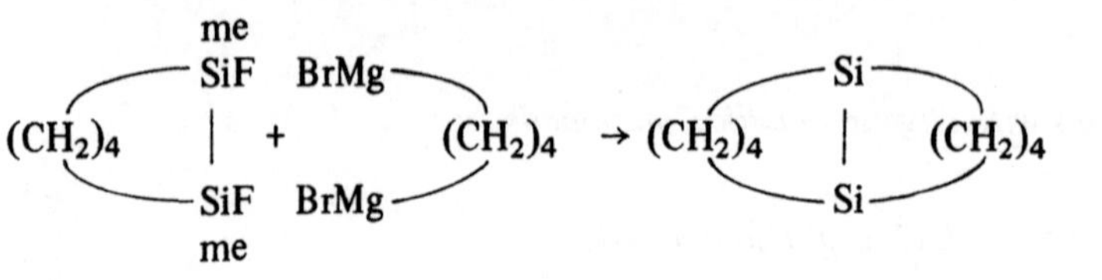

622, 625)

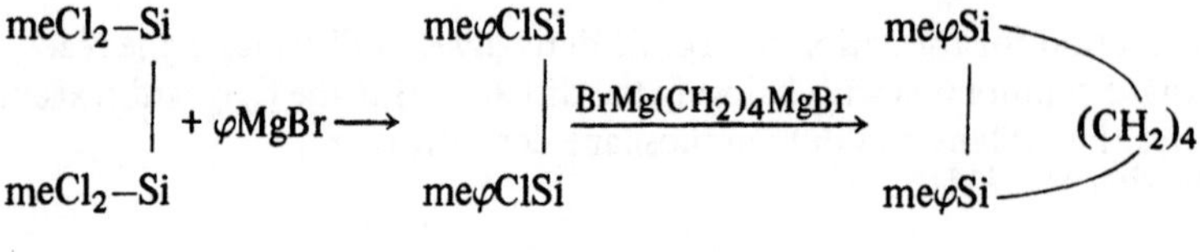

623)

$$me_3SiC_6H_4MgBr \xrightarrow{me_3SiSime_2Cl} me_3SiOC_6H_4\text{–}Sime_2\text{–}Sime_3$$

$$BrMgC_6H_4\text{–}C(me)\langle{}^{O-CH_2}_{O-CH_2}\rangle \xrightarrow[2.)\ H^+]{1.)\ Si_2me_5Cl} Si_2me_5C_6H_4COCH_3$$

481)

$$ClCH_2(Sime_2)_2Cl \xrightarrow{\varphi MgCl} ClCH_2(Sime_2)_2\varphi$$

$$ClCH_2Si(me)(Cl)\text{–}Si(me)_2\text{–}Cl \xrightarrow{\varphi MgCl} ClCH_2Si(me,\varphi)Sime_2\varphi$$

621)

$$Si_2me_5Cl + allyl\ MgBr \longrightarrow Si_2me_5\ allyl$$

$$Si_2me_5Cl + vi\ MgBr \longrightarrow Si_2me_5vi$$

$$Si_2me_5Cl + \varphi MgBr \longrightarrow Si_2me_5\varphi$$

$$Si_2me_5CH_2MgCl + allyl\ Br \longrightarrow Si_2me_5butenyl$$

$$Si_2me_5CH_2MgCl + CO_2 \longrightarrow Si_2me_5CH_2COOH$$

86)

$$et_2meSiSime_2CH_2MgCl + HgCl_2 \longrightarrow et_2meSiSime_2CH_2HgCl$$

$$me_3SiCH_2MgCl + Si_2me_5CH_2HgCl \longrightarrow me_2SiCH_2HgCH_2Si_2me_5$$

$$me_3SiCH_2MgCl + et_2meSi(me_2)SiCH_2HgCl \longrightarrow me_3SiCH_2HgCH_2Si(me_2)Siet_2me$$

$$meMgBr + me_3SiSime_2CH_2HgCl \longrightarrow Si_2me_5CH_2HgCH_3$$

$$meMgBr + et_2meSiSime_2CH_2HgCl \longrightarrow et_2me(Sime_2)CH_2HgCH_3$$

357)

4.2.1.2. Reactions with Organoalkali Compounds

The reactions of organoalkali compounds are similar to those of Grignard compounds. A cleavage of the Si–Si bond sometimes occurs[159]. For example, pentafluorphenyl lithium reacts with Si_2Cl_6 giving only monosilane derivatives[666], but partly chlorinated disilanes like $Clme_5Si_2$ or $Cl_2me_4Si_2$ give the expected disilanes $\varphi^Fme_5Si_2$ and $\varphi^Fme_4Si_2$, respectively[666]. Other examples of reactions investigated are:

$$Si_2Cl_6 + 6\ Na\varphi \longrightarrow Si_2\varphi_6 + 6\ NaCl$$ 578)

$$Si_2Cl_6 + 6\ NaCH_2\varphi \longrightarrow Si_2(CH_2\varphi)_6 + 6\ NaCl$$ 578)

$$Si_2\varphi_5Cl + \varphi Na \longrightarrow Si_2\varphi_6 + LiCl$$ 191)

$$R_3SiSiCl_3 + 3\ LiR' \longrightarrow R_3SiSiR'_3$$
$$R_3SiSiRCl_2 + 2\ LiR' \longrightarrow R_3SiSiRR'_2 \qquad 225)$$
$$R_3SiSiR_2Cl + LiR' \longrightarrow R_3SiSiR_2R'.$$

Metal-exchange reactions are observed, giving unexpected products as with alkali metals[55]:

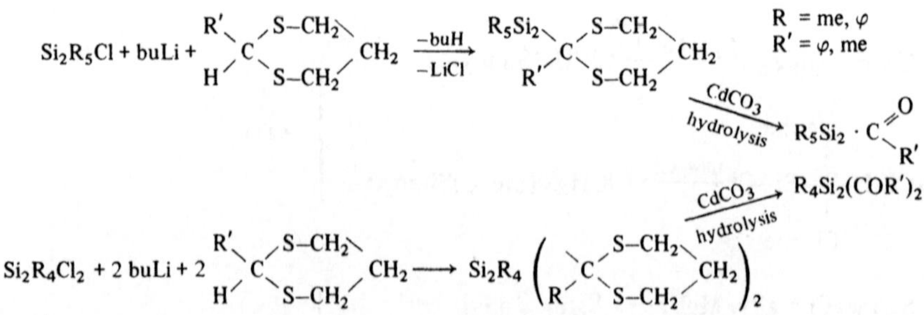

$$CH_2Cl_2 + n\ buLi + ClSi_2me_5 \longrightarrow Cl_2CHSi_2me_5 \qquad 504)$$

Unexpected reaction products are also due to steric hindrance. Gilman and Smart[215] found sym-tetra-o-tolyl-dihydroxydisilane after hydrolysis of a reaction product of Si_2Cl_6 and o-tolyl-Li.

The Si–H bond as well as the Si-halogen bond reacts with organoalkali compounds and this fact can be used formation of new derivatives:

$$Si_2H_6 + Lime \longrightarrow Si_2H_5me + LiH \qquad 44, 234)$$
$$Si_2H_5me + Lime \longrightarrow meH_2SiSiH_2me + LiH. \qquad 234)$$

In certain experiments, alkali or alkaline-earth elements are used in the metallic state. With a Wurtz-like synthesis, new substituents can be inserted at the Si–Si bond:

$$me_3SiSime_2Cl + ClCH_2OCH_3 \xrightarrow[THF]{Mg} me_3SiSiCH_2OCH_3 \qquad 362)$$

$$BrCH_2Cl + ClSi_2me_5 + Li \longrightarrow ClCH_2Si_2me_5 \qquad 504)$$

but sometimes cleavage of the Si–Si bond occurs. Thus, Si_2Cl_6 does not react with sodium and an alkyl halide to the expected hexaalkyldisilane[571], and oligosilanes react with lithium to mono- and disilane lithium derivatives[685] (see p. 49).

4.2.2. Reactions With Halides and Halogens

Hexaalkyldisilanes can be cleaved very easily by free halogens[367, 616]. The cleavage becomes slower[691] with hexaaryldisilanes, and hexahalodisilanes cleave only at high temperature. Cleavage seems to become more difficult with increasing electronegativity of the substituents. ICl also cleaves Si_2me_6 to form me_3SiI, me_3SiCl and I_2[512].

Bromine radicals are able to cleave the Si–Si bond, as demonstrated by Hosomi and Sakurai for the reaction of alkyl- or phenyldisilanes with dibromoethane in the presence of dibenzoyl peroxide[300]:

$$\begin{array}{l} Br\cdot + R'_3SiSiR_3 \longrightarrow BrSiR'_3 + R_3Si\cdot \\ R_3Si\cdot + Br(CH_2)_2Br \longrightarrow R_3SiBr + \cdot CH_2CH_2Br \\ \cdot CH_2CH_2Br \longrightarrow CH_2{=}CH_2 + Br\cdot \\ \hline R'_3SiSiR_3 + BrCH_2CH_2Br \rightarrow R'_3SiBr + R_3SiBr + CH_2{=}CH_2. \end{array}$$

Methylated tri- and tetrasilanes will be cleaved by bromine or iodine to monohalodisilanes[610]. The cleavage of the last Si–Si bond does not occur. The same products were obtained from the halogenation of the corresponding hydrogen compound:

$$\left.\begin{array}{l} Si_3me_8 \\ Si_4me_{10} \end{array}\right\} \xrightarrow{Br_2,\, I_2} Si_2me_5Br(I) \xleftarrow[Br_2(I_2)]{} Si_2me_5H.$$

Photochemical chlorination of hexamethyldisilane yields chloromethyldisilanes. The halogenation takes place[356] on the methyl group:

$$me_nSi_2Cl_{6-n} \xrightarrow[h\nu]{Cl_2} \left\{\begin{array}{ll} (ClCH_2)me_{n-1}\,Cl_{6-n}Si_2 & ^{365,\,620)} \\ (Cl_2CH)me_{n-1}\,Cl_{6-n}Si_2. & ^{356)} \end{array}\right.$$

Halogenation is generally possible in the presence of peroxides as catalyst (see p. 29) for example:

$$me_3SiSiCl_3 \xrightarrow[SO_2Cl_2]{peroxide} ClCH_2(Sime_2)SiCl_3.$$

In the reaction with trimethylsilylchlorosulfinates an exchange occurs[50]:

$$Si_2me_6 + ClSO_3Sime_3 \longrightarrow me_3SiCl + meSO_3(Sime_2)Sime_3.$$

Organofunctional chlorinated disilanes are also formed in low yields in the reaction[619]:

$$Cl_2meSiSimeCl_2 + 2\,CH_2N_2 \longrightarrow Cl(ClCH_2)meSiSimeCl(ClCH_2) \longrightarrow$$

$$\xrightarrow{meMgBr} (ClCH_2)Si_2me_4(CH_2Cl)$$

$$Si_2me_6 + \text{t-bu OCl} \longrightarrow \begin{cases} Si_2me_5CH_2Cl \\ Si_2me_5CHCl_2 \end{cases} \qquad ^{515)}$$

but not in the chlorination of a methoxy group by BCl_3 because a rearrangement occurs[619]:

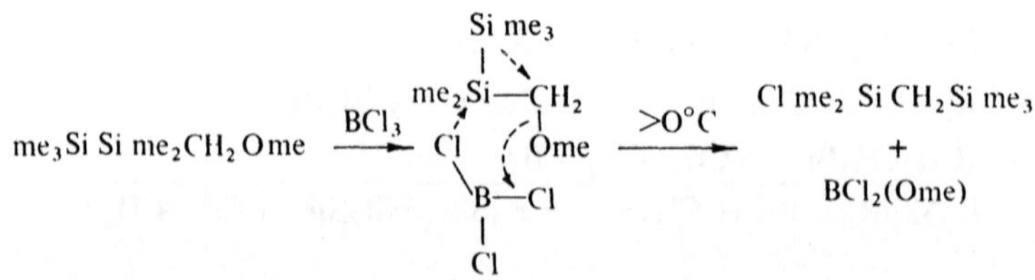

It is remarkable that brominated or iodinated organofunctional disilanes can be formed by reaction of the corresponding methoxy derivatives with triphenyldibromophosphorane[619]:

$$\geq Si{-}Si{-}CH_2Ome + \varphi_3PBr_2 \longrightarrow \geq Si{-}Si{-}CH_2Br + meBr + \varphi_3PO$$

or with KI + H_3PO_4:

$$\geq Si{-}Si{-}CH_2Ome + 2\,KI + 2\,H_3PO_4 \longrightarrow \geq Si{-}SiCH_2I + meI + 2\,KH_2PO_4 + {} + H_2O.$$

A halogen exchange is also possible[619]:

$$(BrCH_2)me_2SiSime_2(CH_2Br) \xrightarrow{2\,NaI} (ICH_2)me_2SiSime_2(CH_2I).$$

Another method of forming iodomethyldisilanes is by the careful iodination of the corresponding thioethers with methyl iodide[489]:

$$me_3SiSime_2CH_2SCH_3 + meI \longrightarrow [me_3SiSime_2CH_2S]^+[me_2I]^-$$

$$\rightleftharpoons me_3SiSime_2CH_2I + me_2S$$

$$(me_2S + meI \longrightarrow [me_3S]I^-).$$

Bromination with N-bromosuccinimide is not possible, because a cleavage of the Si–Si bond occurs[71, 487]. $CuCl_2$ also cleaves the Si–Si bond on chlorination[71].

Mixed disilane derivatives with Si–me and Si–Cl bonds were prepared, usually by the methods of Kumada *et al.* These methods were recently reviewed[368]. Two principal methods are known, firstly methyl exchange by H_2SO_4 and subsequent chlorination or fluorination by NH_4Cl or NH_4F, respectively, and secondly, demethylation and chlorination by HCl or acetyl chloride in presence of $AlCl_3$.

a) The reaction with H_2SO_4/NH_4 *salts:* This reaction yields monochloro-(or monofluoro-)pentamethyldisilane or dichloro-(or difluoro-)tetramethyldisilane[371]:

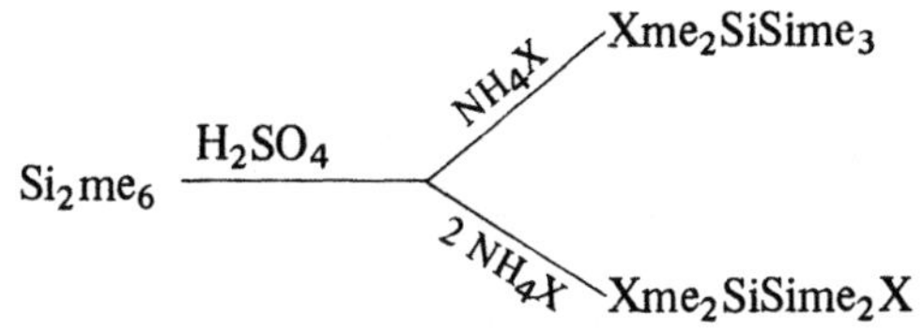

The reaction is also possible with organofunctional methyldisilanes[365]. The intermediate compound after treatment with sulphuric acid can be assumed to be a pentamethyldisilanyl sulfate. This compound was isolated by Kumada and Meada[363]:

$$2\ me_3SiSime_2Cl + H_2SO_4 \longrightarrow (me_3SiSime_2)_2SO_4 + 2\ HCl.$$

In the reaction with sulfuric acid rearrangements may involve a Si–Si cleavage. The following example was shown by Kumada *et al.*[364]:

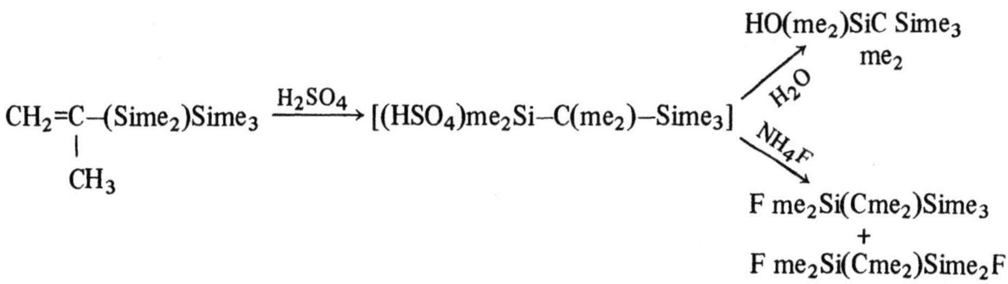

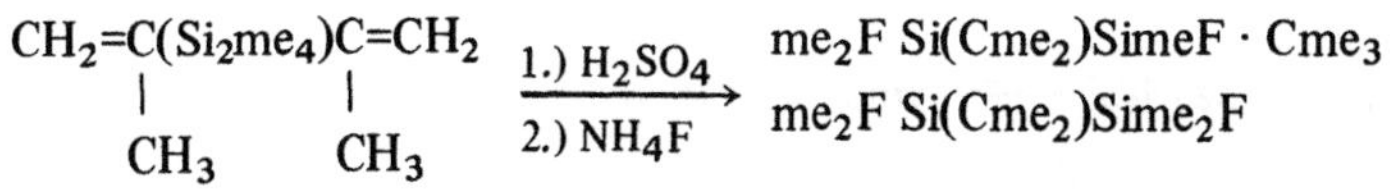

b) Methyl/chlorine exchange with dry HCl *or* CH_3COCl *in the presence of* $AlCl_3$: this reaction yields different chloromethyldisilanes, depending on the reaction temperature:

$$Si_2me_6 \xrightarrow[HCl]{AlCl_3} \left\{ \begin{array}{l} Si_2me_5Cl \\ Si_2me_4Cl_2 \\ Si_2me_3Cl_3 \end{array} \right. \qquad 501)$$

$$Si_2me_6 \xrightarrow[CH_3COCl]{AlCl_3} Si_2me_{6-n}\,Cl_n \quad n = 1, 2, 3, 4$$ [501, 506]

$$\varphi_3SiSime_3 \xrightarrow[HCl]{AlCl_3} Cl_3SiSime_3$$ [620]

$$R.COX + R'_6Si_2 \longrightarrow R.COR' + X\,Si_2R'_5 \quad (X = hal, -OCOR).$$ [151]

Gaseous HCl without $AlCl_3$ does not react with the Si–CH_3 group, which cleaves between 350–700 °C, yielding monosilanes. Partially methylated disilanes proved more stable[514].

Aluminium chloride is also a catalyst for exchange reactions, and can be used for the preparation of chloromethyldisilanes:

$$Si_2me_6 + Clme_2SiSime_2Cl \xrightarrow{AlCl_3} 2\,me_3SiSime_2Cl.$$ [500]

or for skeletal rearrangements of higher polysilanes:

$$me(Sime_2)_n me \xrightarrow{AlCl_3} (me_3Si)_3Sime + (me_3Si)_4Si + (me_3Si)_3Sime_2Sime_3 + \ldots$$
$(n = 4, 5, 6)$. [308]

A similar electrophilic reaction occurs with phenyltrimethylsilane[500], which reacts with a chlorinated methyldisilane faster than with p-methylphenyltrimethylsilane, indicating the electrophilic character of the reaction:

$$Clme_2SiSime_3 + \varphi Sime_3 \longrightarrow \varphi me_2SiSime_3 + ClSime_3.$$ [314, 500]

The last reaction affords a newer and better method of forming mixed-substituent methyl/halogen disilanes by the reaction of trimethylchlorosilane with hexamethyldisilane in presence of $AlCl_3$ as a catalyst[314].

$$Si_2me_6 + 2\,me_3SiX \xrightarrow{AlX_3} Xme_2SiSime_2X \quad X = Cl, Br$$

$$Si_2me_6 + me_3SiX \xrightarrow{AlX_3} Xme_2SiSime_3 \quad X = Cl, I.$$

The reaction is not limited to disilanes; trisilanes may also be used[314].

Aluminium halides are not only able to split methyl groups, but to undergo other reactions, too. The chloromethyldisilane derivatives can be further reacted by rearrangement[365]. The products depend on the reaction conditions. Kumada *et al.*[356] found that, under mild conditions (reaction temperature 70–80 °C), only migration of the trimethylsilyl group from silicon to carbon takes place:

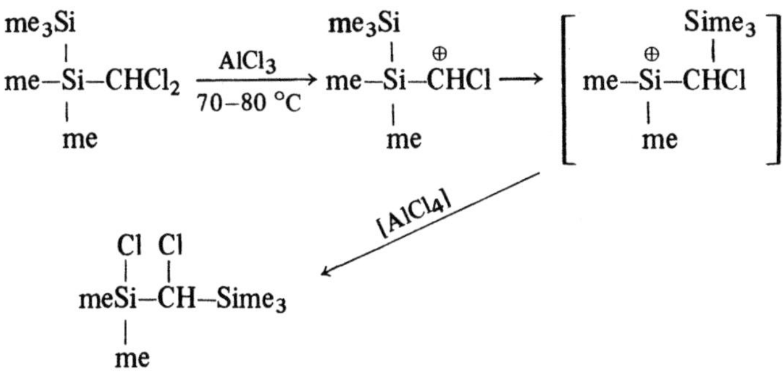

At higher temperature, and employing a large amount of catalyst, a second rearrangement was effected:

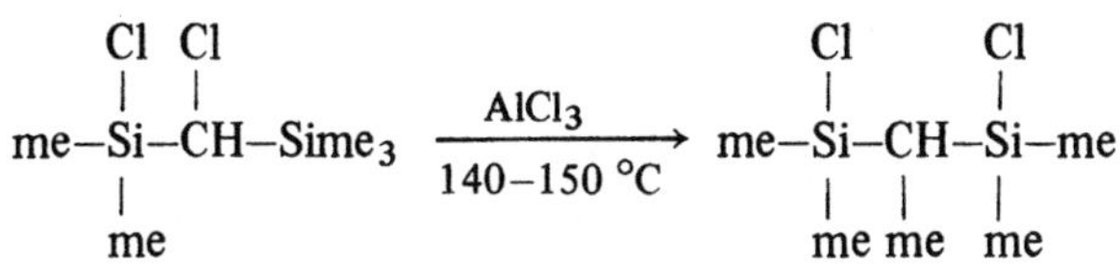

These rearrangements are also possible with halomethyldisilane derivatives with direct Si–Cl bonds. The greater the total number of Si–Cl bonds in the molecule, the less readily the rearrangements occur[620)].

Some data are available on the mechanism of these rearrangements, which also take place with monosilanes[49, 595, 671)]. In the case of disilanes, Kumada *et al.*[620)] postulated a mechanism involving an initially slow rate-determining ionization of the carbon-halogen bond, where the migrating group plays a minor (if any) role, with a subsequent or synchronous, fast nucleophilic attack by the halide ion on the silicon atom followed by migration of the silyl group from silicon to carbon.

Other Lewis basis like BF_3 are not able to cleave the Si–Si bond:

$$(Si_2me_5)_2O \xrightarrow{BF_3} Si_2me_5F \qquad ^{404)}$$

$$(Si_2me_5)_2NH \xrightarrow{BF_3} Si_2me_5F \qquad ^{404)}$$

$$(Si_2H_5)_2NH \xrightarrow{BF_3} Si_2H_5F. \qquad ^{2)}$$

4.2.3. Substituent Exchange Reactions Catalyzed by Peroxides

The hydrogen atom of an alkyl group or of a Si–H bond can be captured by organic radicals such as benzoyl peroxide, etc. The disilanyl radicals formed in this way are readily halogenated.

Some reactions are known for preparation of halogenated disilanes by the following mechanism:

$$2\,Si_2me_6 \xrightarrow[(\varphi COO)_2]{SO_2Cl_2} 2\,me_3SiCl + me_3SiSime_2CH_2Cl$$

$$Si_2me_5H \begin{array}{l} +\varphi H \\ +\varphi CH_3 \end{array} \xrightarrow{+\,t\text{-}bu_2O_2} \begin{array}{l} Si_2me_5\varphi \\ o,m,p\text{-}meC_6H_4Si_2me_5 + \varphi CH_2Si_2me_5 + me_3C{-}Si_2me_5. \end{array} \qquad 482)$$

Other similar substitution reactions are possible:

$$\left.\begin{array}{l} Si_2me_5H + 1\text{-octene} \xrightarrow{t\text{-}bu_2O_2} Si_2me_5\text{-n-octyl} \\ Si_2me_4H_2 + 1\text{-octene} \xrightarrow{t\text{-}bu_2O_2} Si_2me_4(\text{n-octyl})H \end{array}\right| \; 644)$$

Di-t-butylperoxide or benzoylperoxide reacts with Si_2me_6 in CCl_4 primarily producing $Si_2me_5CH_2$ radicals which further react with chlorine atoms from the solvent[492]. In an inert solvent, the radicals are able to react with for example $CH_2{=}CH.\ Si_2me_5$, or, in the absence of another reagent, between themselves to form dimers:

$$Si_2me_6 \xrightarrow[\text{or } \varphi\cdot CO(OOH)]{\text{di-t-bu-peroxide}} Si_2me_5CH_2\cdot \xrightarrow{CCl_4} Si_2me_5Cl$$

$$Si_2me_5H + CH_2{=}CHSi_2me_5 \xrightarrow{\text{di-t-bu-peroxide}} Si_2me_5CH_2CH_2Si_2me_5.$$

The dimeric compound was also prepared by an independent synthesis[492]. Similar investigations had been carried out before with ethyl derivatives[686]. Kinetic measurements are available for the radical H-splitting by t-butylperoxide[486].

Perbenzoic acid reacts with organic disilanes by oxidation to the corresponding disiloxanes[488] in a second-order reaction. A substitution on the phenyl group of the phenylpentamethyl-disilanes increases the nucleophilic reaction rate of the Si–Si oxidation and increases $p_\pi{-}d_\pi$ conjugation from the phenyl-group to the silicon.

4.2.4. Preparation of Derivatives of Disilanes by Hydrogenation

Like all silicon-halogen bonds, halogenated disilanes are hydrogenated by complex hydrides. A series of new disilane derivatives are thereby synthesized:

$$\left.\begin{array}{l} Si_2me_5Cl \\ Si_2me_4Cl_2 \\ (RO)_2meSiSime_2(OR) \\ Cl_2meSiSimeCl_2 \\ (RO)_2meSiSime(OR)_2 \\ Si_2\varphi_5Cl \end{array}\right| \xrightarrow{LiAlH_4} \left|\begin{array}{ll} Si_2me_5H & 644) \\ Si_2me_4H_2 & 358) \\ H_2meSiSime_2H & \\ H_2meSiSimeH_2 & 644) \\ H_2meSiSimeH_2 & \\ Si_2\varphi_5H & 681) \end{array}\right.$$

Hydrogenation with hydrogen in presence of Raney nickel can be effected without cleavage of the Si–Si bond[330]:

$$\varphi_6Si_2 \xrightarrow[110-200^\circ]{H_2(Ni)} (hexyl)_6Si_2\,.$$

However, cleavages of the Si–Si bond were observed during hydrogenation in presence of an Al_2O_3/KOH catalyst. Kumada *et al.*[615] found cleavage and formation of Si–H bonds with partially chlorinated methyldisilanes.

4.2.5. Nucleophilic Substitution Reactions

A large series of compounds can be prepared by nucleophilic exchange. The method is limited by the susceptibility of the Si–Si bond to nucleophilic attack.

4.2.5.1. Hydrolysis

One of the most frequent reactions in silicon chemistry is the hydrolysis of Si-halogen bonds:

$$-\overset{|}{\underset{|}{Si}}-Cl + HOH \longrightarrow -\overset{|}{\underset{|}{Si}}-OH + HCl.$$

This reaction can be carried out with disilanes[575].

$$2\ Si_2me_5Cl \xrightarrow{H_2O} (Si_2me_5)_2O + 2\ HCl \qquad 96)$$

but the reaction medium must not become alkaline or the Si–Si bond will be destroyed:

$$-\overset{|}{\underset{|}{Si}}-\overset{|}{\underset{|}{Si}}- + 2\ HOH \xrightarrow{OH'} 2\ -\overset{|}{\underset{|}{Si}}-OH + H_2$$

Detailed investigations of the hydrolysis of disilanes to form heterocyclic compounds were carried out by Kumada *et al.*[360] (for more details, see Chapter 9.2: Heterocyclic Silanes).

Hydrolysis of the Si–H bond in alkaline solution in presence of a Si–Si bond destroys both linkages. Separation of the two reactions is not possible. This was investigated by Schott and Langecker for the case of the hydrolysis of tetraalkyl and tetraaryldisilanes[565]. Substitution on the phenyl group changes the reaction rate. A relation between the rate constants and the Hammet constants was observed, but for linearity the Hammet constants have to be reduced to take into account a mesomeric factor. Special σ_{Si} constants are therefore necessary (there are similar differences in the σ constants of other Si compounds[16, 628]).

4.2.5.2. *Solvolysis by Alcohols*

Some papers have described the preparation of alkoxy disilanes by the reaction of halogenated disilanes with alcohols. Several examples are given in the following equations:

$$Si_2Cl_6 \xrightarrow{ROH} Si_2Cl_n(OR)_{6-n} \quad n = 0\text{–}6 \quad \begin{cases} R = me & ^{438)} \\ \phantom{R = {}}et & ^{4, 388, 416, 438, 439)} \\ \phantom{R = {}}i\text{-}pr & ^{439, 612)} \end{cases}$$

R = et, CHme$_2$, nC$_3$H$_7$, –CH(me)et [60] (for Si_2Cl_6 and for the following reaction)

$$Si_2me_5Cl \xrightarrow{ROH} Si_2me_5OR$$

$$Si_2me_{6-n}Cl_n \xrightarrow{ROH} Si_2(OR)_n me_{6-n} \quad n = 1, 2, 3, 4 \qquad ^{260)}$$

$$\varphi_5Si_2Cl \xrightarrow{ROH} Si_2\varphi_5(OR) \qquad R = et, CH_2\varphi \qquad ^{191)}$$

The equations indicate that solvolysis is also possible. However, in most cases the reactions give only small yields. Many side reactions, cleavage of the Si–Si bond, rearrangements, and condensation reactions to polymers take place.

For example, methanol also reacts with the Si–H bond:

$$H(Sime_2)_nH \xrightarrow[\text{trace of buLi}]{CH_3OH} meO(Sime_2)_nOme + H_2 \quad n = 2, 3, 4. \qquad ^{21)}$$

Sometimes it seems to be better to use alkoxy derivatives like the ester of nitrous acid instead of free alcohols for the preparation of alkoxydisilanes:

$$Si_2Cl_6 + 6\ NO(Ome) \longrightarrow Si_2(Ome)_6 + 6\ NOCl \qquad ^{674)}$$

or acetic anhydride:

$$Si_2Cl_6 + 6\ acOac \longrightarrow Si_2(Oac)_6 + 6\ AcCl. \qquad ^{376)}$$

With acetate ions in acetic acid, a two-step reaction occurs, firstly a substitution to (acetylmethyl)-pentamethyldisilane and secondly a rearrangement to acetoxypentamethyldisilylmethane:

$$Si_2me_5CH_2Cl + me \cdot COONa/meCOOH \longrightarrow Si_2me_5CH_2OOC \cdot me \longrightarrow$$

$$\longrightarrow \underset{\displaystyle O \cdot CO \cdot me}{\underset{|}{me_2Si}}\!-\!CH_2\!-\!Sime_3 \qquad ^{362)}$$

Sulphur compounds can be similarly prepared with mercaptans and mercaptides:

$$Si_2me_5CH_2Cl \xrightarrow[\text{or HX}]{C_2H_5SNa/C_2H_5SH} Si_2me_5CH_2Set \qquad ^{362)}$$

$$(\text{with } X = Set, S\varphi)$$

Cleavages of the Si–Si bond mostly occur in alkaline solutions in the presence of alcoholates[641]. Detailed investigations were carried out by Kumada and by Atwell and Weyenburg. Kumada *et al.*[614] investigated the cleavage rate of methylalkoxydisilanes by sodium alcoholate in alcohol. It is initially directly proportional to the concentration of the disilane and of the alcoholate ions. The splitting is facilitated, when the disilane is substituted by some electrophilic groups. Symmetric disilanes have a greater activation energy and entropy change in the reaction than do asymmetric substituted disilanes.

Atwell and Weyenburg[21] investigated in detail the base-catalyzed rearrangement of methoxymethyldisilanes, previously discovered by Ryan[477]

$$(meO)_2meSi\!-\!Sime(Ome)_2 \longrightarrow meSi(Ome)_3 + \text{„polysilanes"}$$

$$(meO)me_2Si\!-\!Sime_2(Ome) \xrightarrow[100^\circ]{LiOme} me_2Si(Ome)_2 + meO(Sime_2)_3Ome$$

$$(meO)(Sime_2)_5Ome + me_2Si(Ome)_2 \xleftarrow[180^\circ]{LiOme}$$

The reaction mechanism will be assumed to be a four-center process:

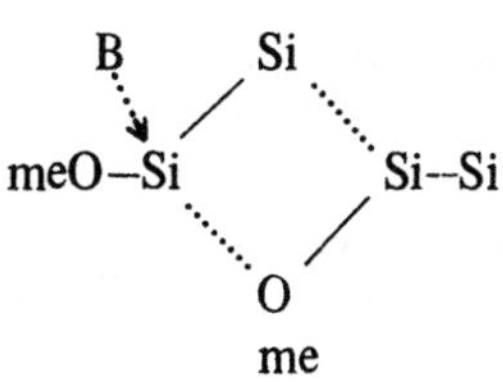

In the nucleophilic splitting of chloromethyldisilanes both splitting and rearrangement reactions were observed[356, 362, 369, 621] and three possible mechanisms postulated:

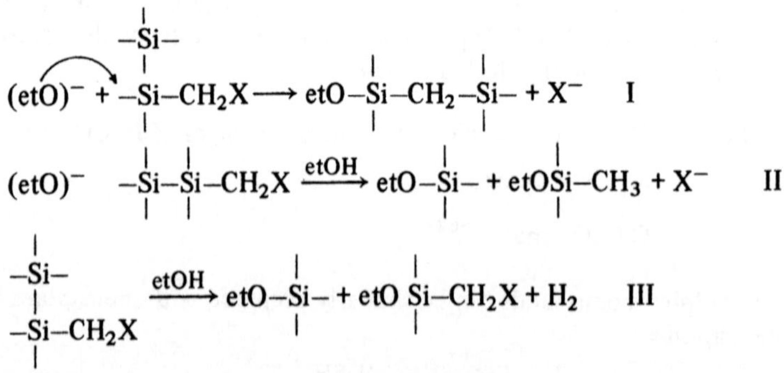

Methylated chloromethyldisilanes react according to I (2/3) and II (1/3). Phenylation on the α-silicon atom leads to the pathway II reaction.

A plausible mechanism involving formation of a silylene anion is given by the authors:

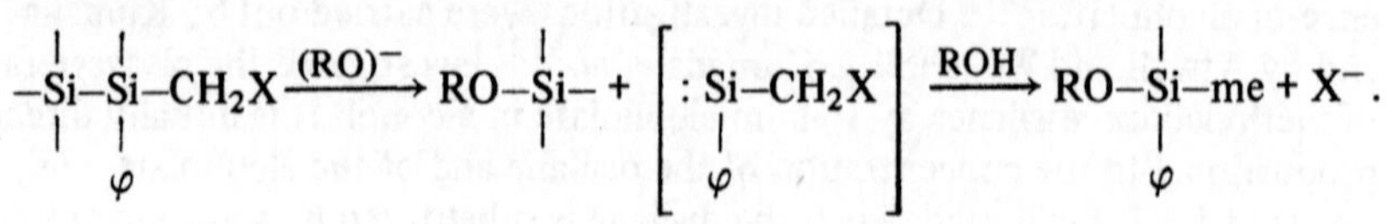

The preference of the cleavage (II) with α-phenylated disilanes is due to the higher stability of a phenyl-substituted silylene anion. The alkoxy ion primarily attacks the β silicon atom.

Chloromethyldisilanes with partial chlorination directly at the silicon atom also react by route I or II, except for the derivative $Cl_3SiSime_2(CH_2Cl)$, which is decomposed with of evolution hydrogen. The following mechanisms were discussed:

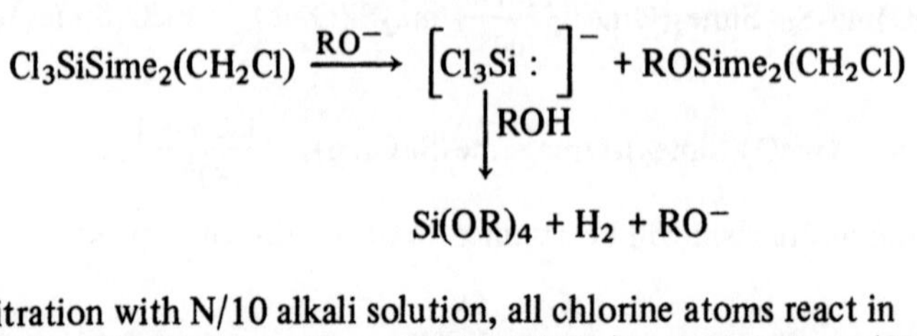

In titration with N/10 alkali solution, all chlorine atoms react in $Cl_2meSiSiCl_2(CH_2Cl)$ but in $Cl_2meSiSiCl_2(CHCl_2)$ and $Cl_3SiSime_2(CH_2Cl)$ only the silicon-halogen bond is acid enough for reaction with the alkali. β-elimination takes place under these reaction conditions (reaction is only allowed to proceed to the neutral point) with $Cl_2meSiSiCl_2(CH_2Cl)$. Such β-elimination is also believed to occur with β-chloroethylchlorosilane:

$$HO^- + -\overset{|}{\underset{|}{Si}}-CH_2-CH_2-Cl \longrightarrow HO-\overset{|}{\underset{|}{Si}}- + CH_2{=}CH_2 + Cl^-.$$

For the second compound, a splitting of the whole $CHCl_2$ group has to be assumed:

$$Cl_2meSiSiCl_2(CHCl_2) \xrightarrow[ROH]{RO^-} meSi(OR)_3 + (RO)_4Si + CH_2Cl_2 + H_2$$

The reaction of the third compound was described previously. Ketones[174], aminoxides[490] and NO_2[532] are also able to cleave the Si–Si bond by a nucleophilic reaction yielding siloxanes.

A completely different reaction takes place if Si_2me_6 is reacted with meONa in presence of hexamethylphosphoric acid triamide[498] (see p. 21). Trisilane thiols readily undergo a radical-induced skeletal rearrangement:

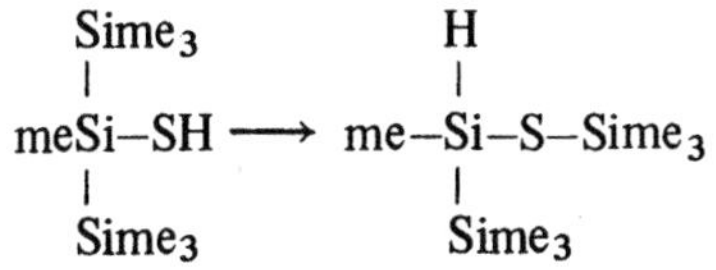

providing examples of 1, 2 shifts of the silyl and disilanyl groups which occur in solution under mild conditions[302].

4.2.5.3. Solvolysis by Ammonia and Amines

Pentamethylchlorodisilane reacts with ammonia to bis-(pentamethyldisilanyl)-amine[404]:

$$2\ Si_2me_5Cl + NH_3 \longrightarrow (Si_2me_5)_2NH + 2\ HCl.$$

Si_2Cl_6, too, certainly reacts with amines. With secondary amines, not only is the hexasubstituted compound obtained but partially substituted compounds can also be isolated.

$$Si_2Cl_6 + 12\ HNme_2 \longrightarrow Si_2(Nme_2)_6 + 6\ HNme_2 \cdot HCl$$ [675]

$$Si_2me_{6-n}Cl_n + 2\,n\ HNme_2 \longrightarrow Si_2me_{6-n}(Nme_2)_n + HNme_2 \cdot HCl.$$ [270]

In a side reaction, nucleophilic Si–Si cleavage sometimes takes place[52]:

$$Si_2Cl_6 + 7\ HNet_2 \longrightarrow Cl_2Si(Net_2)_2 + ClHSi(Net_2)_2 + 3\ HNet_2 \cdot HCl.$$

Reactions with organofunctional disilanes are also known[362]:

$$Si_2me_5CH_2Cl + et_2NH \longrightarrow Si_2me_5CH_2(Net_2)$$
$$Si_2me_5CH_2Cl + \varphi NH_2 \longrightarrow Si_2me_5CH_2(N\varphi H)$$

Ammonia reacts to polymeric compounds (see p. 93).

Monobromodisilane reacts with dimethylamine to form the expected mono(dimethylamino)disilane[2] which was found to be more stable than the corresponding monosilane derivative. In this reaction additional formation of the bis(dimethylamino)disilane was observed. This can be explained according to the equation:

$$2\ Si_2H_5(Nme_2) \longrightarrow Si_2H_4(Nme_2)_2 + Si_2H_6 .$$

The reaction is probably caused by an excess of me_2NH, which can form an addition product with the α silicon atom:

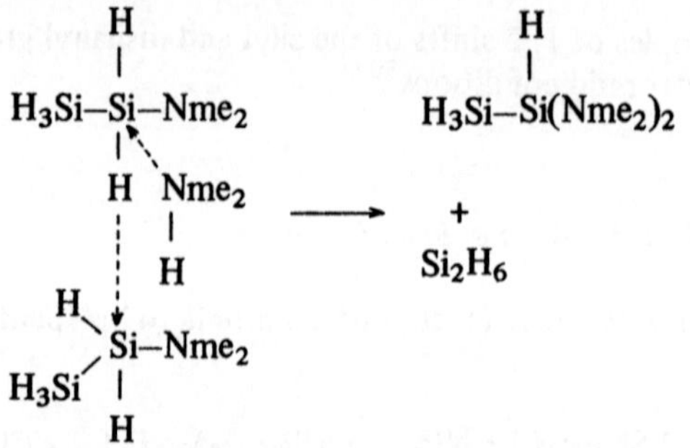

The disubstituted product is more stable than the monosubstituted one and remains stable for several days at room temperature.

Alkali amides cleave the Si–Si bond, forming compounds with Si–N bonds:

$$Si_2me_6 + bu_2NLi \longrightarrow me_3SiNbu_2 . \qquad ^{98)}$$

Derivatives of hydrazine are also known, produced by the following reactions:

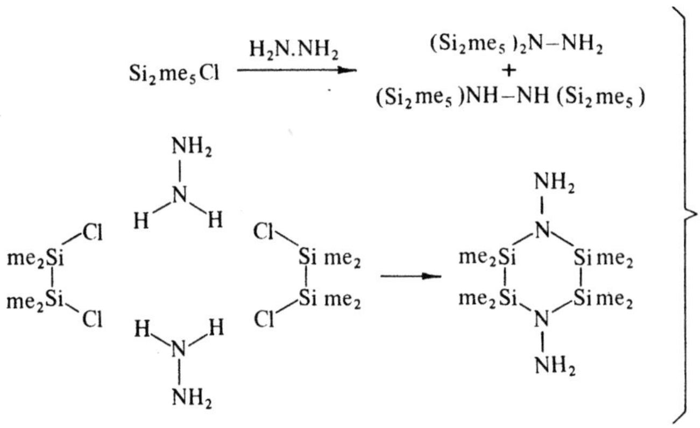

89)

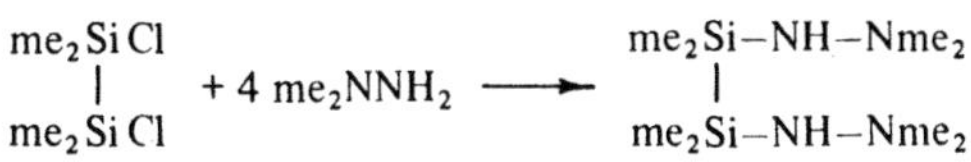

661)

Nonsubstituted hydrazine reacts to form cyclic compounds. With substituted hydrazine derivatives, no cyclization takes place.

Si_2Cl_6 reacts with ammonium halides giving cleavage of the Si–Si bond and formation of monosilanes and polymeric material[678]. Mixed alkylhalogenodisilanes also react, but not hexaalkyldisilanes.

The reactions of trialkylamines with hexachlorodisilane were discussed in Chapter 3 (see p. 11).

A nucleophilic cleavage also takes place with phosphine oxide:

$$Si_2Cl_6 + O{=}PR_1R_2R_3 \longrightarrow :PR_1R_2R_3 + Si_2OCl_6 \qquad 429)$$

4.2.5.4. Reactions with Cyanides and Cyanates

Chloroalkyldisilanes react with cyanides and cyanates to form the expected derivatives. Whether the derivatives are in the n or the i form was investigated in some cases.

The i form was characterized by IR measurements of a cyanate obtained in the reaction:

$$Si_2me_5Cl + AgCNO \longrightarrow Si_2me_5NCO + AgCl. \qquad 404)$$

The cyanide Si_2H_5CN obtained in the reaction of Si_2H_5I with AgCN[96] seems to be a mixture of n and i forms with strong association. A Si–Si cleavage occurs on warming pentamethyldisilanylcyanide reacts above 175° to trimethylcyanide and methylcyanopolysilanes[643]. The corresponding trisilanes react in a similar way:

$$n\ me_3Si\text{–}(Sime_2)_2CN \longrightarrow (n-1)me_3SiCN + me_3Si(Sime_2)_{2n}CN.$$

A four-center mechanism will be assumed in this as in similar reactions.

Cyanates are also known:

$$Si_2R_4Cl_2 + AgCNO \longrightarrow Si_2R_4(NCO)_2. \quad R = \varphi, me \qquad ^{449)}$$

The i cyanate formed reacts with dry piperidine without cleaving the Si–Si bond:

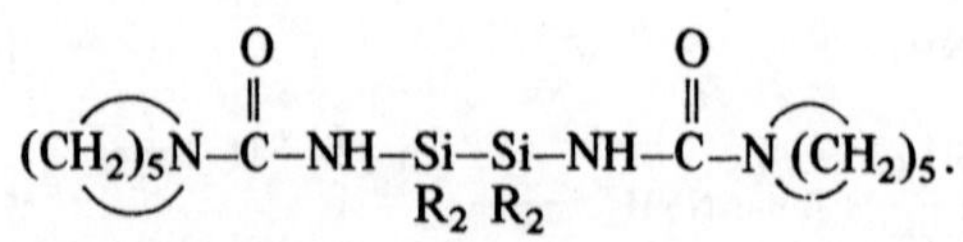

It should be possible to incorporate the tetraorganodisilane unit into polymeric systems by reaction through these isocyanate linkages.

Organofunctional reactions are also known:

$$Si_2me_5CH_2Cl \xrightarrow{(CN)_2} Si_2me_5CH_2CN. \qquad ^{362)}$$

The cyanide formed reacts with alcohol in the presence of a trace of cyanide ion:

$$Si_2me_5CH_2CN + C_2H_5OH \xrightarrow{CN^-} Si_2me_5OC_2H_5 + meCN. \qquad ^{362)}$$

4.2.5.5. Noble Metal-Catalyzed Reactions of the Si–Si bond

Disilanes and trisilanes with olefinic substituents (such as vinyl- and allyl-disilanes) react in the presence of $H_2PtCl_6 \cdot 6\,H_2O$ undergoing a Si–Si cleavage[692]. A second reaction sometimes occurs so that simple cleavage products are not obtained. For example, vinylpentamethyldisilane reacts with the cleavage product trimethylsilane, giving an ethane derivative:

$$me_5Si_2CH{=}CH_2 + HX \xrightarrow{[Pt]} me_3SiH + Xme_2SiCH{=}CH_2$$

$$me_5Si_2CH{=}CH_2 + me_3SiH \xrightarrow{[Pt]} me_5Si_2CH_2CH_2Si_2me_3.$$

The catalytic effect of platinum salts can be explained by the intermediate formation of an olefin-metal π complex in which the electron density of the α silicon atom is decreased so nucleophilic attack becomes easier. Other nucleophilic cleavages such as alcoholysis are facilitated and side reactions also occur.

$$me_3SiSime_2CH{=}CH_2 + etOH \xrightarrow{H_2PtCl_6 \cdot 6\,H_2O} me_5Si_2CH_2CH_2Sime_3 + etOme_2SiCH_2CH_2Sime_3$$

$$me_3SiSime_2CH{=}CH_2 + etOH \xrightarrow{PdCl_2} me_3SiOet + CH_2{=}CHSime_2Oet + etOme_2Siet + me_5Si_2et.$$

Under the same reaction conditions, alkyl disilanes without olefinic groups do not react.

Recent investigations of Kumada *et al.*[693, 694] show that the catalytic action of platinum compounds leads to rearrangements of pentamethyldisilane and tetramethyldisilane to higher silanes:

$$me_3SiSime_2H \xrightarrow{[Pt]} \Sigma\, me(Sime_2)_nH \quad n = 1\text{–}6$$

$$Hme_2SiSime_2H \xrightarrow{[Pt]} \Sigma\, H(Sime_2)H \quad n = 1\text{–}6$$

The reaction mechanism can be explained by the formation of a dimethylsilylene intermediate which is able to undergo an insertion reaction into a Si–H bond (for more about this type of reaction, see Chapter 7: Silylenes). The formation of the silylene can be explained by an α elimination from the π metal-complexed disilane. The presence of the silylene has been proved by a capture reaction with diphenylacetylene, when the expected 1,1,4,4-tetramethyl-1,4-disilacyclohexadiene could be isolated.

4.3. Stereospecificity of Disilane Derivatives

Kumada, Ishikawa and Tamoa[617, 618, 623] reported the preparation of disilane derivatives with cis-trans isomerism. A mixture of cis-trans 1,2-diphenyl-1,2-dimethyl-1,2-disilacyclohexane was obtained from the following reaction:

$$Cl_2me\,Si{-}Si\,me\,Cl_2 + 2\,p{-}XC_6H_4MgCl \longrightarrow [p{-}X\,C_6H_4meClSi]_2 \quad X{=}H,Cl$$

I

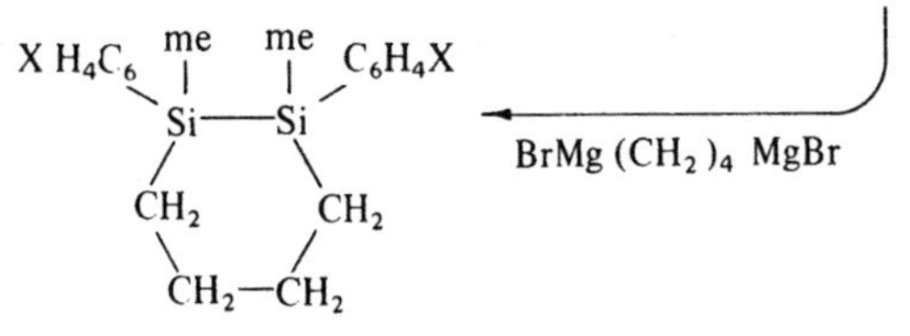

II

The optically active derivative I was founded in a mixture of d-, 1- and meso-forms. The cis-trans isomers of II could be separated by careful fractional distillation or by gas chromatography. The configuration of the chlorophenyl derivatives could be estimated by dipole measurements. The chlorine atoms were exchanged with hydrogen by Grignard reaction and hydrolysis, so the configuration of the phenyl compounds was also determined. The cis and trans forms differ in their physicochemical constants and in their NMR spectra.

The described cleavage of organic groups from the silicon atom by interaction with sulphuric acid enables exchange of the phenyl group for chlorine or fluorine:

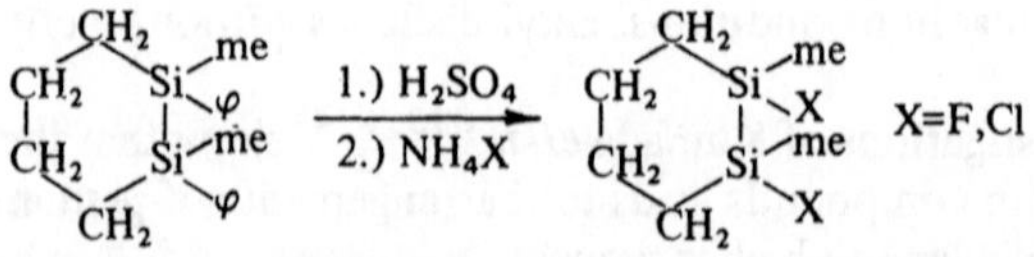

In this reaction a mixture of cis and trans isomers always occurs irrespective of the starting material (cis, trans, or a mixture). Separation into isomers may also be effected by gas chromatography. The different forms were identified from the phenyl derivatives obtained by reaction with phenylmagnesium chloride.

The formation of both isomers after the fluorodephenylation seems to occur not only in the interaction with sulphuric acid, but also with alcohol or moisture. The authors postulated a mechanism similar to the racemization of 1-naphthylphenylmethylfluorosilane, which included a pentacovalent intermediate state:

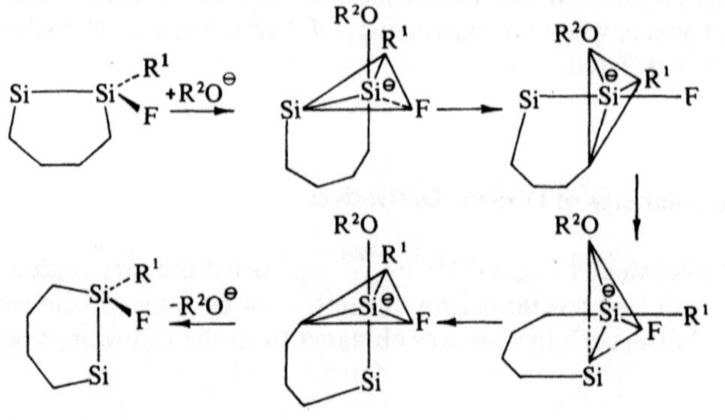

With perbenzoic acid or bis(trimethylsilyl)peroxide, oxidation of the Si–Si bond occurs:

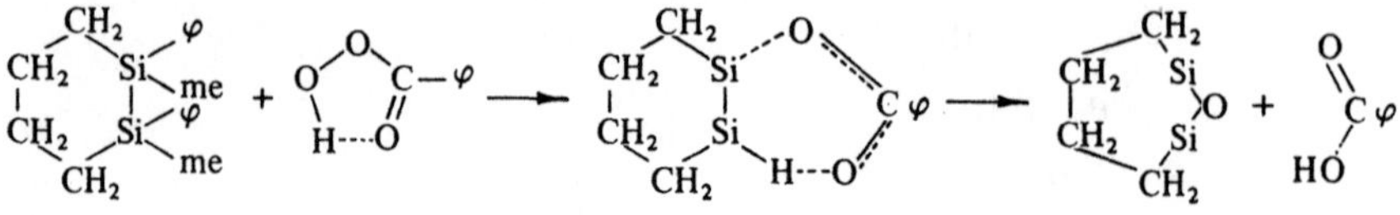

This reaction is stereospecific; a trans starting compound yields only a trans product.

The difluoro compound I reacts with $BrMg(CH_2)_4$ MgBr to form cis- or trans-9,10-dimethyl-9,10-disiladecaline[622, 625] (II):

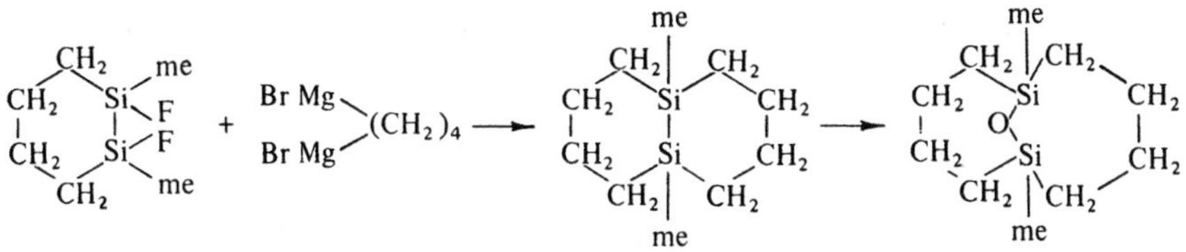

I II III

This cis form of I reacts only to the cis form of II, but the trans form of I reacts to give a mixture of cis and trans isomers due to a ring straining in the trans starting compound.

With peroxybenzoic acid or bis(trimethylsilyl)peroxide, II can be oxidized to cis-1,6-dimethyl-II-oxa-1,6-disila-[4,4,1]-undecane (III). This compound can be obtained only in the cis form since in the trans form the ring straining is too great.

We shall now discuss the stereochemistry of an intramolecularly fully bridged disilane. Kumada, Tamoa and Noro[624] recently reported the following reaction:

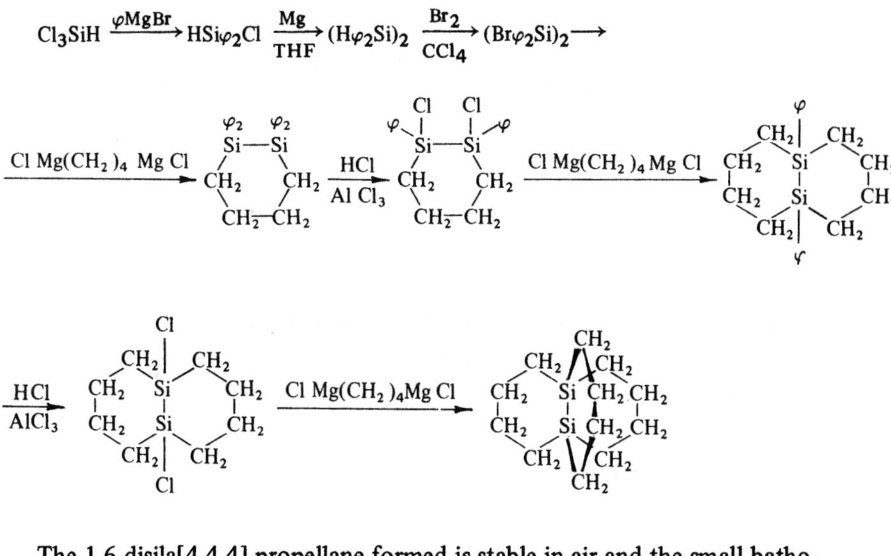

The 1,6-disila[4,4,4]-propellane formed is stable in air and the small bathochromic shift relative to hexamethyldisilane can be explained by ring straining.

4.4. Optically Active Derivatives of Disilanes

Optically active monosilanes were investigated in detail, mostly by Sommer and coworkers[526]. Investigations of disilanes are very rare, with only a few published papers.

The first investigation was done by Kipping[381]. He synthesized an asymmetric disilane:

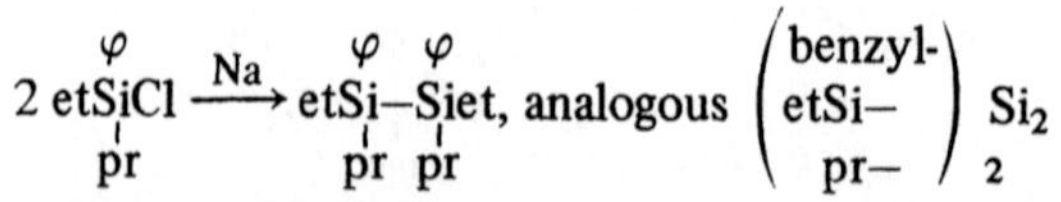

but he was not able to separate the optical isomers, although he formed a corresponding sulfo derivative $[Si(et)(pr)(CH_2C_6H_4SO_3H)]_2$ and the (–)(menthyl)salt.

In the studies of optically active monosilanes, a second optically active disilane was prepared in 1965[527]:

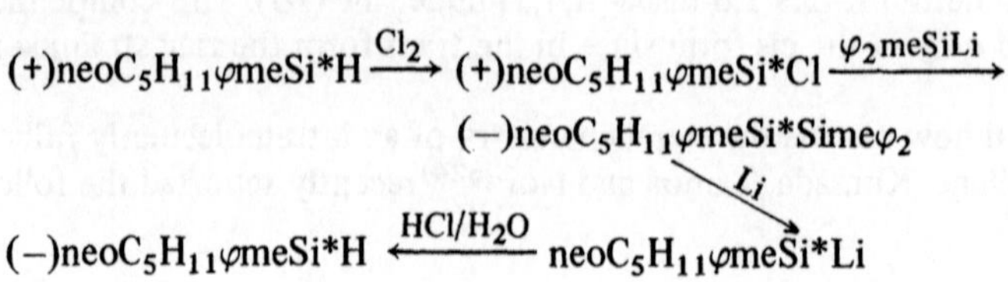

This is the first stereospecific cleavage of a Si–Si bond by an alkali metal. Functional optically active disilanes were first prepared in 1967 by Sommer *et al.*[528, 529], and the optically active isomers were successfully separated.

Resolution of the 1,2,2,2-Tetraphenyl-1-methyldisilane System

(±)Ph_3SiSi*O(–)Men(Ph)(Me)
I, II, [α]D –34.1° (*c* 2.82, pentane)

(+)Ph_3SiSi*O(–)Men(Ph)(Me)	(–)Ph_3SiSi*O(–)Men(Ph)(Me)
I, [α]D +20.0°, mp 90–91°	II (1), [α]D –55.5°
$LiAlH_4$ ↓	$LiAlH_4$ ↓
(–)Ph_2SiSi*H(Ph)(Me)	(+)Ph_3SiSi*H(Ph)(Me)
III, [α]D –14.3°, mp 66–67°	IV (III), [α]D +7.7°, mp 68–69°
Cl_2 ↓	Cl_2 ↓
(+)Ph_3SiSi*Cl(Ph)(Me)	(–)Ph_3SiSi*Cl(Ph)(Me)
V, [α]D +12.0°, mp 86–88°	VI (V), [α]D –6.5°, mp 87–94°

fractional crystallization

$$(-)Ph_3SiSi^*Cl(Ph)(Me) \qquad (\pm)Ph_3SiSi^*Cl(Ph)(Me)$$

$$\text{VI, } [\alpha]_D\ -12.0°,\ \text{mp } 86\text{–}88° \qquad 1{:}1\ \text{V:VI, } [\alpha]_D\ 0°,\ \text{mp } 94\text{–}95°$$

In this investigation, the Walden rearrangement with disilanes was first observed:

$$\underset{\text{III}}{(-)Ph_3SiSi^*H(Ph)(Me)} \xrightarrow{Cl_2} \underset{\text{V}}{(+)Ph_3SiSi^*Cl(Ph)(Me)}$$

$$\text{III} \xleftarrow{LiAlH_4} \text{VI} \qquad \text{V} \xrightarrow{LiAlH_4} \text{IV}$$

$$\underset{\text{VI}}{(-)Ph_3SiSi^*Cl(Ph)(Me)} \xleftarrow{Cl_2} \underset{\text{IV}}{(+)Ph_3SiSi^*H(Ph)(Me)}$$

A probable mechanism was discussed.

A recent detailed investigation[530] was carried out on the UV optical rotatory dispersion and the circular dichroism of the compounds $\varphi_3SiSi\varphi(me)X$ (with X = H, F, OH, OCH_3). A Cotton effect was seen and discussed in relation to the structure.

4.5. Pyrolysis Reactions of Disilanes

Decompositions and rearrangements, also synthetic reactions are observed.

Most details about the pyrolysis of hexamethyldisilane are known, including some estimations of the Si–Si bond energy[23, 90, 91, 101, 102] (see also p. 52). At atmospheric pressure, decomposition starts at about 600 °C and yields mainly $me_3SiCH_2Sime_2H$[513] and me_3SiH[494].

A mechanism involving radicals was postulated:

$$me_6Si_2 \longrightarrow 2\cdot Sime_3 \xrightarrow{Si_2me_6} me_3SiH + me_3SiSime_2CH_2\cdot$$

$$me_3SiSime_2CH_2\cdot \longrightarrow me_3Si\text{–}CH_2\text{–}Sime_2\cdot$$

$$me_3SiCH_2Sime_2H + me_3SiSime_2CH_2\cdot \xleftarrow{Si_2me_6} me_3Si\text{–}CH_2\text{–}Sime_2\cdot$$

At a lower pressure (0.2–0.8 mm Hg) and temperatures of 523–555 °C other reaction products are observed[101]:

$(me_2SiCH_2)_n$, 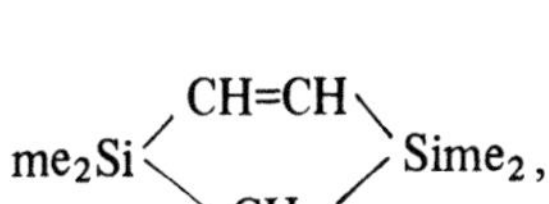, $Si_2me_5CH_2Si_2me_5$, $Si_2me_5CH_2Sime_3$.

Perhaps changed reaction conditions result in a different mechanism and other products, because a repetition of the work of Kumada gave the same results[494, 513]. In the pyrolysis of Si_2me_5et under similar conditions (0.1–0.7 mm Hg, 545 °C) the monosilane derivatives Si_2me_6 and $Si_2me_4et_2$ were observed. A radical mechanism was discussed[102]:

$$2\ me_3SiSime_2et \rightleftharpoons 2\ me_3Si\cdot + 2\ me_2etSi\cdot \rightleftharpoons me_6Si_2 + me_4et_2Si_2$$

Si_2me_6, Si_2me_5Cl or $Si_2me_4Cl_2$ react at 500–600 °C in the presence of chlorobenzene to compounds with the Si–CH_2–Si group and methylchlorosilanes[85].

Hexaethyldisilane is formed from et_4Si or $et_3Si\varphi$ on heating to 350–360 °C in presence of hyperbaric hydrogen[305]. SiH_4 reacts with C_2H_2 at 450–510 °C forming higher silanes $etSi_2H_5$, Si_2H_6 etc., but only in small yields[156a, 670]. Si_2D_6 reacts with $meSiH_3$ to form meH_2SiSiD_2H[132] and Si_2me_6 was formed on heating of $Sime_4$[88]. Other examples are known.

Radical mechanisms are postulated for all these pyrolysis reactions. Modification of the radical mechanism by the reaction conditions, would explain the variable results.

Another major possibility would be a mechanism involving silylenes. This was postulated in 1969 by Kumada and Sakurai[485] for the pyrolysis of pentamethyldisilane. Methylpolysilanes are formed in an insertion reaction into silylenes:

$$me_3SiSime_2H \rightleftharpoons me_3SiH + me_2Si:$$

$$me_2Si: + me_3SiSime_2H \rightleftharpoons me_3Si(Sime_2)_nH. \qquad n = 3, 4, 5$$

Similar results were found by Baird *et al.*[22]:

$$meH_2SiSiH_2me \xrightarrow{400\ °C} meSiH_3 + 1,2,3\text{–}me_3Si_3H_5, \text{ etc.}$$

$$H_2meSiSiH_3 \xrightarrow{342-360\ °C} meSiH_3 + SiH_4 + meSi_3H_7 + me_2Si_3H_6, \text{ etc.}$$

and Skell and Goldstein[518] in a reaction of a vapor stream of me_2SiCl_2 over K/Na alloy:

$$me_2SiCl_2 \xrightarrow{K/Na} me_2Si: + me_3SiH \longrightarrow me_3SiSime_2H.$$

Mechanisms of silylene insertion reactions are known for the methoxymethylpolysilanes. Detailed investigations are reported in Chapter 7: Silylenes.

The factors important in:

a) a radical mechanism

$$Si_2R_6 \rightleftharpoons 2 \cdot SiR_3$$

b) a silylene mechanism

$$Si_2R_6 \rightleftharpoons SiR_4 + :SiR_2$$

were shown by Davidson[100]. In relation to thermodynamic studies he postulated that:

1. If the activation energy for the formation of silylenes is smaller than the activation energy for the formation of radicals, a silylene mechanism occurs;
2. if the activation energy of the insertion reaction is high and the activation energy of the formation of the radicals is smaller than the activation energy of the formation of the silylenes, a dissociation to radicals occurs first.

Since an insertion reaction into alkyl or aryl silicon bonds seems to be difficult, a radical mechanism usually takes place in the pyrolysis of such compounds. On the other hand, at Si–H, Si–Hal, Si–OR and Si–N bonds, a silylene mechanism occurs because the insertion reaction has only a small activation energy.

4.6. Synthesis of Disilanes by a Silent Electrical Discharge

The reaction elicited by silent electrical discharge have a similar mechanism to pyrolysis, though investigations have been less complete.

SiH_4 reacts with me_2O under electrical discharge to me. $SiH_2 \cdot SiH_3$ and other unidentified products[3]. Methylsilane, $meSiH_3$ reacts specifically to 1,2-dimethyldisilane, potentially via a radical or a silylene mechanism[7]:

$$\underline{2\ meSiH_3 \longrightarrow H_2 + Si_2me_2H_4}$$

$$meSiH_3 \begin{cases} H\cdot + me\dot{Si}H_2 \xrightarrow{meSiH_3} meSiH_2{-}SiH_2me + H\cdot \\ H_2 + meHSi: \xrightarrow{meSiH_3} meSiH_2{-}SiH_2me. \end{cases}$$

As seen in an earlier chapter, the silylene mechanism is preferred in presence of a Si–H bond. However, observations on pyrolysis need not apply to reactions, induced by electrical discharge.

Other reactions by silent electrical discharge which have been investigated are:

$$\left.\begin{array}{l} 2\ me_2SiH_2 \xrightarrow{\not{z}} me_2SiH{-}SiHme_2 + H_2 \\ 2\ me_3SiH \xrightarrow{\not{z}} me_6Si_2 \\ meSiH_3 \xrightarrow{\not{z}} meSi_2H_5 + H_2 \end{array}\right| \quad 7)$$

$$SiH_4 + me_2SiH_2 \xrightarrow{\text{↯}} \left. \begin{array}{c} meH_2SiSiH_2me \\ + \\ me_2HSiSiH_3 \end{array} \right| \quad ^{234)}$$

4.7. Preparations of Disilanes by Photochemical Reactions

After a describing the conversion of methylsilanes to disilanes in the presence of mercury under photochemical conditions[430)], Urry and Reedy reported the photochemical preparation of methylchlorodisilanes[467)]:

$$meCl_2SiH \xrightarrow[(Hg)]{h\nu} \left\{ \begin{array}{l} Cl_2meSiSimeCl_2 \\ Cl_2meSiSime_2Cl \end{array} \right.$$

A subsequent chlorination of the resulting mixture yields only 1,1,2,2-tetrachlorodimethyldisilane. Preparation of Si_2Cl_6 from $SiHCl_3$ or 1,1,2,2-tetramethyldichlorodisilane from dimethylchlorosilane was similarly possible.

4.8. Adducts of Disilanes

Although adducts between the Lewis acid silicon and sundry Lewis bases are important as intermediates in the reactions of Si–Si bonds, very few stable adducts are known.

Si_2Cl_6 reacts with trimethylamine to form two adducts, $Si_2Cl_6 \cdot Nme_3$ and $Si_2Cl_6 \cdot 2\ Nme_3$[328)]. These adducts are starting materials for some rearrangements leading to higher silicon halides (see p. 11).

Schott *et al.*[559)] have prepared some adducts of Si_2Cl_6 with urea: $Si_2Cl_6 \cdot 10\ CO(NH_2)_2$, $Si_2Cl_6 \cdot 10\ CO(NHme)_2(NH_2)$ and $Si_2Cl_6 \cdot NH\varphi$. In these compounds, adduct formation seems to involve bonds between the amine hydrogen and the chlorine of the disilane.

Kummer *et al.*[374)] found addition compounds of Si_2Cl_6 and $Si_2Cl_4me_2$ with bipyridyl. They postulate a six-coordinated compound.

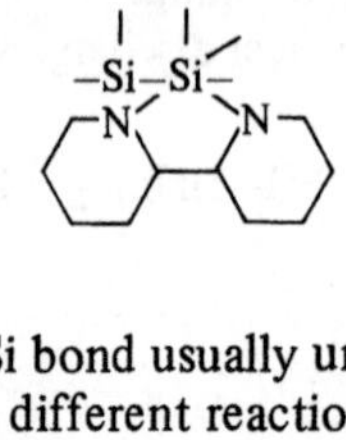

The Lewis base-sensitive Si–Si bond usually undergoes cleavage in reaction with such bases. Indeed, with different reaction conditions, a cleavage reaction is observed and adducts with the monosilanes and the silylene are formed, respectively[372)]:

$$Si_2Cl_6 + 3\ bipy \xrightarrow[25°]{THF} SiCl_2 \cdot 2\ bipy + SiCl_4 \cdot bipy \qquad 372)$$
$$\downarrow$$
$$(SiCl_2)_n$$

Si_2Cl_6 + py → Cl_3Si 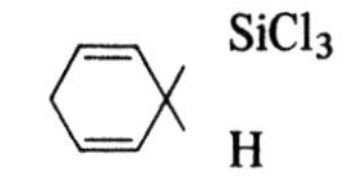 373)

Similar results were obtained by Wannagat *et al.*[659].

5. Higher Chain and Branched Organopolysilanes

Many kinds of longer chain and branched silanes with organic substituents are known. The principal reactions of the Si–Si bonding in these substances correspond with those which have been discussed for disilanes.

The methods of preparation are the usual ones for formation of Si–Si bonds. In several cases one can start with the easily accessible cycloorganosilane derivatives and obtain the corresponding chains by cleavage. Octaphenylcyclotetrasilane is particularly suitable for these cleavage reactions, in which a series of tetrasilane derivatives can be prepared and still longer chain derivatives synthesized through further coupling reactions. A detailed scheme of the octaphenylcyclotetrasilane reaction sequence is given on p. 79.

The five-membered ring is much more stable, though it can be cleaved by chlorinating media, resulting in shorter chains. Cleavage by Li also gives a partial cleavage of the pentasilane chain[222].

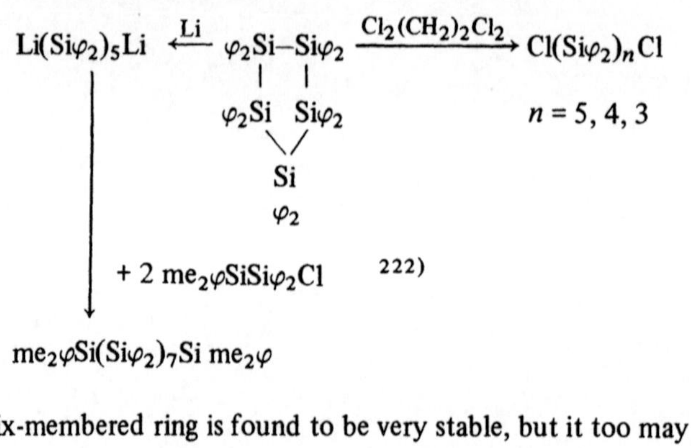

The six-membered ring is found to be very stable, but it too may be cleaved:

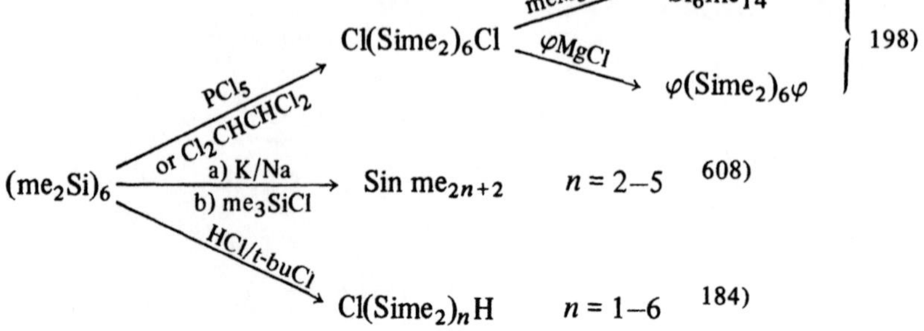

Wurtz synthesis and similar coupling mechanisms have also been used in the preparation of higher silane derivates:

$$2\ \mathrm{et_3SiSiet_2Br} + 2\ \mathrm{Na} \longrightarrow \mathrm{et_3Si(Siet_2)_2Siet_3} + 2\ \mathrm{NaBr} \qquad ^{466)}$$

$$\left.\begin{array}{l} \mathrm{me_3SiSime_2Cl} \xrightarrow{\mathrm{Na/K}} \mathrm{Si_4me_{10}} \\ \mathrm{me_3SiSime_2Cl} + \mathrm{ClSime_3} \xrightarrow{\mathrm{Na/K}} \mathrm{Si_3me_8} \end{array}\right\} \quad ^{505,\ 607,\ 679)}$$

$$\varphi_3\mathrm{SiK} + \varphi_2\mathrm{SiCl_2} \xrightarrow{\mathrm{Na}} \mathrm{Si_3}\varphi_8 \qquad ^{227)}$$

$$\mathrm{Li(Si}\varphi_2)_4\mathrm{Li} \xrightarrow{2\ \mathrm{ClSi}\varphi_2\mathrm{H}} \mathrm{H(Si}\varphi_2)_6\mathrm{H} \qquad ^{195)}$$

Kumada *et al.*[359] prepared linear methylpolysilanes up to $n = 12$ by Wurtz synthesis from different chloromethyldisilanes and found by investigation of the physical constants that the melting points of the silanes up to $n = 7$ do not rise continuously but follow in a zigzag course; since the polysilanes with an odd number of silicon atoms melt at a lower temperature, than those with an even number of silicon atoms, though this effect is less pronounced with higher n.

Irradiation with UV (2537 Å) causes decomposition of longer chains (n = 4–6) into $\mathrm{Si_3me_8}$ and polymer products, presumably by a silylene mechanism[312].

Partly methylated higher di-tri- and tetrasilanes are created by insertion of Si atoms (produced by electron bombardment on hot silicon electrodes) into the Si–H bonding[440, 519]:

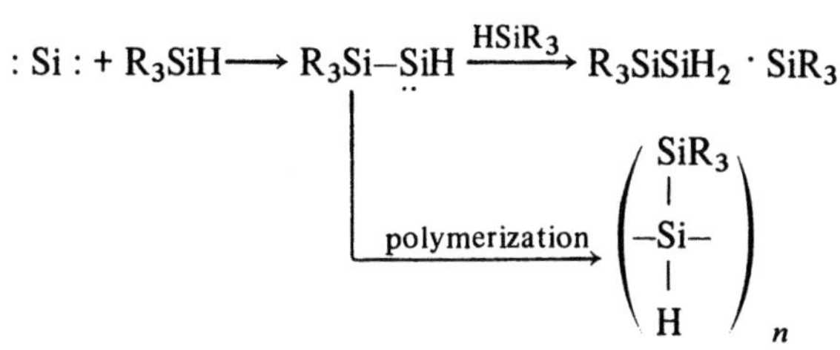

Arylpolysilanes are cleaved by alkali metals to smaller fragments:

$$\left.\begin{array}{l} \mathrm{Si_3}\varphi_8 \xrightarrow{\mathrm{Li}} \varphi_3\mathrm{SiLi} + \varphi_3\mathrm{SiSi}\varphi_2\mathrm{Li} \\ \mathrm{Si_4}\varphi_{10} \xrightarrow{\mathrm{Li}} \varphi_3\mathrm{SiLi} + \varphi_3\mathrm{SiSi}\varphi_2\mathrm{Li} + \varphi_3\mathrm{Si(Si}\varphi_2)\mathrm{Si}\varphi_2\mathrm{Li.} \end{array}\right. \quad ^{685)}$$

Alkylpolysilanes are cleaved slowly with alkali metals, as shown by Stolberg[609]. Complicated conversion and transmetalation reactions also occur, since higher chain (up to $n = 7$) and cyclic methylsilanes appear after interruption of the reaction by addition of $\mathrm{me_3SiCl}$. A silylene insertion mechanism also seems to be active here.

The first branched organopolysilane was found by Gilman[685] (1959) in the reaction:

$$\varphi_3SiK + HSiCl_3 \longrightarrow (\varphi_3Si)_3SiH.$$

Exploiting this line of thought, a neopentane structure could be produced with Si:

$$ClSime_3 + Li + SiCl_4 \longrightarrow (me_3Si)_4Si. \qquad ^{126,\,218)}$$

This compound reacts with ph_3SiLi or methyl lithium to the corresponding lithium derivative, which affords a series of products:

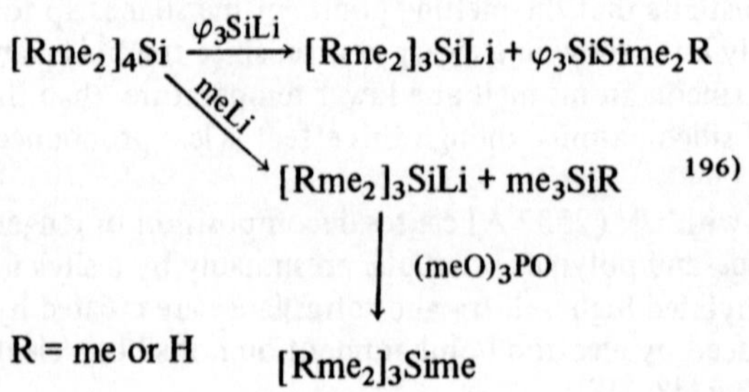

Further derivatives which have been prepared are tetrakis(dimethylphenylsilyl)silane[217] and pentachlorophenyl-tris(trimethylsilyl)silane[213]. The chlorination of a methyl group by treatment with $CH_3COCl/AlCl_3$[507] gives $Si(Sime_2CH_2Cl)_4$.

Tris(trimethylsilyl)silane can be prepared from the Li compound by acid hydrolysis as reported by Gilman[193]:

$$(me_3Si)_3SiLi + (H_3O)^+ \longrightarrow Li^+ + H_2O + (me_3Si)_3SiH$$

or can be obtained directly (Bürger[66]):

$$HSiCl_3 + 6\,Li + 3\,me_3SiCl \xrightarrow[-15^\circ]{THF} 6\,LiCl + (me_3Si)_3SiH.$$

Tris(dimethylphenylsilyl)silane originates from me_2SiLi and Cl_3SiH[183]. The tertiary Si–H group in these substances has been investigated in detail (UV, IR, NMR) and was found very reactive and slightly oxidizable. The chloro compound initially formed undergoes further reactions:

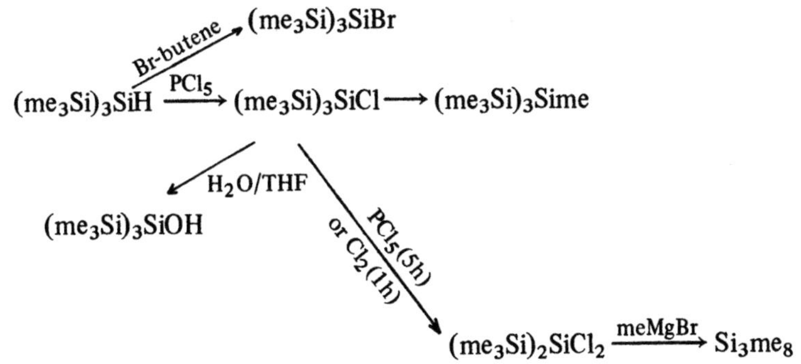

PCl_5 and such halohydrocarbons as CCl_4, $CHCl_3$, C_4H_9Br, $CHBr_3$ and CHI_3 [67] react with the Si–H bond (see also p. 8).

The fluorine derivative is obtained from the hydrogen compound by fluorination with SbF_3 [67].

By coupling $(me_3Si)_3SiLi$ with $(me_3Si)_3SiCl$ or removing Li with $BrCH_2CH_2Br$, hexakis(trimethylsilyl)silane is formed:

$$(me_3Si)_3SiLi + ClSi(Sime_3)_3 \longrightarrow (me_3Si)_3SiSi(Sime_3)_3 .$$

The effects of additional bonding on the Si–Si framework in reactions of tris(trimethylsilyl)silyl lithium have been investigated by Gilman and Smith [219] and no essential differences from triphenylsilyl lithium could be found.

Spectral characteristics of this class of compounds have been measured by Bürger *et al.* [64–67]. From NMR and vibration spectroscopy studies with normal coordinate analysis and calculation of the force constants, a clear weakening of the Si–Si and Si–H bonding seems indicated, due to the inductive effect of the me_3Si group. Some of the measured data are extreme (νSiH in $(me_3Si)_3SiH$: 2050 cm^{-1}, $J^{29}SiH$ = 155 c/s).

There are no differences between the physicochemical parameters of the methyl and the ethyl derivatives, indicating that only steric effects produce differences in their chemical properties.

6. Physicochemical Properties of the Si–Si Bond

The main interest of physicochemical properties in compounds with Si–Si bonds are the bond strength, the influence of electronic effects on the Si–Si bond, and the mechanism of possible additional bond forces. Several values have been reported[61, 122, 521] for the bond energy of the Si–Si bond. For Si_2me_6, Hess, Lampe and Sommer[284, 285] obtained a value as high as 86 kcal/mol by electron impact studies, while Connor *et al.*[90, 91] determined values of 58 and 49 kcal/mol by the toluene carrier technique. A higher bond energy was calculated by Steele and Stone[593] with 81,3 kcal/mol for Si_2H_6 and 85 kcal/mol for Si_2Cl_6, but another finding was 48,9 kcal/mol[28, 236] with a similar figure for Si_3H_8[237] of 50.4 kcal/mol. The changes in the bond energy of Si–H (0.3) and Si–Si (1.5 kcal/mol) on passing from Si_2H_6 to Si_3H_8 are much smaller than the corresponding changes from C_2H_6 to C_3H_8. This could be because the large size of the silicon atom results in smaller interactions between non-bonding atoms.

Different bond-energy values were obtained from pyrolysis and mass-spectroscopic measurements[125, 478–480]. Davidson and Steve[23, 101, 102] found a value of 67 kcal/mol for pyrolysis of me_6Si_2 and Lappert *et al.*[394] found 68 kcal/mol, which appeared to be the most reliable figure. There are similar significant discrepancies in the values for the heats of formation for Si_2H_6[140, 236, 306, 458] and Si_2me_6[306], determined by calorimetric methods. Figures for the standard enthalpy of Si_2me_6 are: -96 ± 1.7 kcal/mol and -87.0 ± 2.4[394].

All these values indicate that the Si–Si bond is a relatively weak bond, as is confirmed by tritium recoil reactions with Si_2me_6[99].

The strength of the Si–Si bond has an important effect on the mass spectra of higher silanes. Fragmentation of disilane and its derivatives[478, 479] shows the existence of a relatively stable „siliconium ion" $R_1R_2R_3Si^+$. Its stability in those fragments depends on the type of substituents. Sakurai *et al.*[491] found that the stabilizing effect of a phenyl group was about ten times that of a methyl group, despite the appreciable stabilization energy of a triphenylsilyl cation, but only half that observed for φ_3C^+[158]. The low stabilization energy of φ_3Si^+ indicates a very small resonance between the phenyl groups and silicon. Mass spectra of cyclic compounds $(Si\varphi_2)_n$ $n = 4, 5, 6$ show the expected molecular ions and fragment series resulting from the stepwise cleavage of neutral and positively charged φ_2 Si fragments[393]. The appearance of fragment groups is particularly clear in the mass spectra of Si_5H_{10}, including Si_nH_x fragments with $n = 5, 4, 3, 2, 1$ and $x = 2n$ or less and the expected isotope fragments[255].

The Si–Si bond distance has been calculated by various methods. The results are summarized in Table 1.

Table 1. Si–Si bond distance

Compound	Si–Si [Å]	Method	Ref.	Rotation barrier [kcal/mol]
Si_2H_6	–	ED	239)	1
	–	V	444)	1.22
	2.32	ED	53)	
Si_2me_6	2.34	ED	26)	
	2.34	ED	54)	
Si_2Cl_6	2.34	ED	536)	
	2.294	ED	424)	1
	2.32	ED	53)	0.65
	2.34	ED	695)	
Si_2H_5F	2.332	MW	94)	
$H_3SiSiHF_2$			401)	0.85
H_3SiSiF_3			401)	0.81
$F_3SiSiF_2BF_3$	2.36	ED	83)	2.35
$Si(Sime_3)_4$	2.361	ED	24)	
$Si_4O_2(me)_8$	2.35	X-ray	613)	
$Si_8O_4(me)_8$	2.36	X-ray	286)	
Si metallic	2.352	X-ray	442)	

MW = micro wave
V = vibration
ED = electron diffraction

The results show that the Si–Si bond distance has a value of 2.34 Å in Si_2me_6, but in Si_2Cl_6 it seems a little smaller. In the solid, elemental state, a greater bond length is observed. If these differences are significant, different bonding conditions are inferred. The observed differences in the bond distances would complement the calculated force constants.

Raman and IR spectra of tri-, i-, and n-tetrasilane were measured, but no force constants were calculated.

Hexahalodisilanes were investigated in detail. Some results of vibrational measurements for Si_2Cl_6 are known[63, 330, 332, 423, 596]. For Si_2Cl_6, Si_2Br_6 and Si_2I_6, a normal coordinate analysis was first carried out in 1969 and the force constants f(Si–Si) were calculated by a general valence force field and an iteration method[27]. The resulting data are listed in Table 2. Recent measurements and calculations for Si_2F_6 agree well with these data. An increase in the force constants is observed with increased electronegativity of the halogen atom. Whether the increased strength of the linkage is caused by a (d–d)π interaction produced by the electronegativity and the (d–p)π bonds of the halogen atoms, or by an interaction in the σ bond system of the Si–Si bond, cannot be

Table 2. Force constants of the Si–Si bond

Compound	f_{Si-Si} mdyn/Å	Ref.
Si_2H_6	1.78	554)
	1.73	37)
Si_2F_6	2.4	293)
Si_2Cl_6	2.4	292)
Si_2Br_6	2.1	292)
Si_2I_6	1.9	292)
Si_2me_6	1.70	150)
	1.65	289)
$Si_2(Ome)_6$	2.25	291)
$(me_3Si)_3SiX$	1.44	66)
M_4Si_4 M(Na, K, Ba. .)	1.05	62)

decided on the basis of these calculations. The disilane molecule is generally considered to have a small rotational barrier giving D'_{3h} symmetry. Recent electronic diffraction measurements show a shortening of the expected C . . . . C distance in the trans position which gives rise to a deviation of 10 ± 4^0 from full staggering. It is reasonable to conclude that the molecule is actually staggered, but that the low-energy torsional vibrations about the Si–Si bond give rise to a shrinkage effect[26)].

The barrier height for rotation about the Si–Si bond is usually estimated to be about one kcal/mol, which is lower than the barrier of ethane or digermane. Some detailed data are given in Table 1. Suga *et al.* [533, 534)] found by X-ray analysis and differential thermoanalysis that, at about –50 °C, a transition of the structure of Si_2me_6 occurred; this should be capable of relation to twisting about the Si–Si axis. NMR relaxation and the activation energy were studied during transition[238)], but dipole measurements are rare[331, 443, 523)].

Numerous papers were published on the vibrational spectra of disilane Si_2H_6 [117, 135, 239, 390, 596, 677)]. Pfeiffer and Spangenberg calculated the force constants[444)]. The estimated Si–Si rotation barrier, 1.22 kcal/mol, correlates well with other data, though the force constant of the Si–Si linkage seems a little too high (1.78 mdyn/Å). More recent calculations taking into account anharmonicity gave an estimated 1.73 mdyn/Å[289)]. Measurements with Si_2me_6 [79, 150, 617)] result in the expected low force constants, because there is little or no additional bonding between carbon and silicon (see Table 2).

The resonance spectra of mixed substituted disilanes, such as the series of halomethyldisilanes, were measured several times[245, 248, 611, 642)] and recently investigated in detail[290)]. An increase in the force constant *f*(Si–Si) occurs on passing from Si_2me_6 to Si_2Cl_6 and amounts to about 0.1 mdyn/Å for each added chlorine atom, with a larger final increase to Si_2Cl_6. Similar results are obtained in the series $Si_2(OR)_{6-n}Cl_n$, in which an increase in *f*(Si–Si) also occurs; the highest value reached is 2.25 mdyn/Å for hexamethoxydisilane[291)].

For similar mixed derivatives of disilanes like etO/me, /me, et/me[72] vi/me[124], me/φ[246], CN/me, NCO/me[642] only vibrational spectra were determined.

Calculations of force constants show that the strength of the Si–Si linkage is easily influenced by substituents. The Si–Si force constant is now expected to range from 1.44 to 2.5 mdyn/Å.

Force constant calculations from vibrational spectra are only possible with relatively small molecules, so efforts were made to correlate force constants with other data. Bürger *et al.*[62] found a good correlation with the bond length. The data for halodisilanes do not fit the curve, though with shorter bond distances the correlation improves.

The Si–H bond in a disilane is influenced by the neighboring Si–Si bond, so correlation of the force constant (Si–Si) with the force constant (Si–H) and the corresponding valence resonance was attempted[66, 250]. The correlation in similar compounds is good and allows derivation of the force constant (Si–Si) for known Si–H resonances in the hydrogen derivative of half the disilane molecule.

The electron-acceptor properties of the Si–Si bond and the effect of electron-donating substituents result in a larger inductive effect of the Si_2me_5 group than in $Sime_3$ group. This was observed in IR and Raman investigations of disilanes with olefinic substituents[86]. From NMR measurements it is concluded that the p–me_5Si_2 group in phenyl derivatives has a larger +I effect and also a larger –T effect than the p–$Sime_3$-group[481]. Stabilization of the Si–Si bond by d_π–d_π interaction was postulated on the basis of kinetic measurements of the base-catalyzed hydrolysis of the Si–H bond in polysilanes[77].

The NMR data for most disilane derivatives are known. However, only a small number of studies have been published relating NMR data to the properties of the Si–Si bond. The chemical shift of the protons of Si_2me_6 shows only a very small difference in comparison with $Sime_4$ (Si_2me_6 : 9.963 ± 0.02 ppm[59]), indicating the small influence of the Si–Si bond.

NMR investigations of Si_2H_6 and Si_3H_8[123] showed that the resonance of the silyl protons in Si_3H_8 appears at the lower end of the silylene resonance in contrast to the tendency in the corresponding carbon compound C_3H_8. Furthermore, a difference of about 0.03 ppm in the chemical shift is observed between $^{29}SiH_3$ protons and $^{28}SiH_3$ proton. This difference is unusually large for an isotope effect. The coupling constant of 199 c/s in Si_2H_6[110] is in good accord with recent observations on the new compound Si_5H_{10} (195 c/s)[255].

Van Dyke and MacDiarmid[119] observed the chemical shifts in the directly bonded ^{29}Si–H and the vicinal (H, H′) coupling constants for a series of compouns H_3Si–SiH_2X with X = H, F, Cl, Br, I. O, SiH_3, CH_3, N, or S. The results suggest some interaction between the substituent X and the H_3Si group in H_3Si–SiH_2X and so, they agree with the force-constant calculations for disilane derivatives (see above).

Similar results are given by Brune *et al.*[60] for the investigation of Si_2me_5OR by NMR Brune postulated a d_π–d_π interaction between the two silicon atoms, which also agrees with the calculated force constants. Good accord also exists with the explanations of the color of polymeric silicon compounds (see p. 111).

Linear polymethylsilanes up to Si_{10}, cyclic polymethylsilanes ($n = 5, 6, 7$), and monosubstituted pentamethyldisilanes and heptamethyltrisilanes (O, H, Cl, φ) were investigated by West, Kumada and coworkers[669], who found in systematic NMR investigations that the longer the chain, the lower the field where the chemical shift of the methyl protons appears. The distance of the methyl group shift from the end of the chain increases with chain length. Electronegative substituents decrease the shielding of methyl protons on the silicon atom to which they are attached. The effect decreases in the order $Cl > \varphi > O > H$.

d-Orbital involvement in the Si–Si bond was investigated by Cartledge and Riedel[78]. The observed coupling constants ^{29}Si–H and ^{13}C–H in comparison to benzylsilanes

```
        me me
        |  |
XC6H4 Si–Si–H      (X = H, p–Ome, p–me, m–me, m–F, m–Ome)
        |  |
        me me
```

show good correlation with the Hammet constants. This and the absence of significant differences in the coupling constants of the mono- and disilane derivatives indicates no enhanced transmission such as could be associated with d_π–d_π bonding, p_π–d_π bonding or a difference in polarizability between carbon and silicon. A detailed investigation[320] of Si_2F_6 was carried out, determining ^{19}F NMR the isotope effect of silicon, and coupling constants, but no conclusion was arrived at with respect to the Si–Si bonding system.

Very few papers have been published on ^{29}Si NMR spectroscopy of compounds with Si–Si bonds, namely of $(me_3Si)_3Sime$ and $(me_3Si)_2$[555]. Similar NQR measurements are available for Si_2Cl_6[243] (^{35}Cl NQR in Si_2Cl_6 : 19.29 megacycles, in Si_2OCl_6 : 19.89 megacycles. More data are required to interpret the effect of the Si–Si linkage.

UV data have been obtained for a large number of polysilane derivatives, and the influence of the substituents on the Si–Si bonds discussed. The data were recently summarized[406]. Most of these investigations were carried out by Gilman and coworkers[179, 181–183, 185, 204] and by Kumada *et al.*[495, 499, 502, 503], but also by other authors[170, 244, 260, 375, 456]. First measurements were made by Hague and Prince[240–242].

It is interesting to observe the influence of the electronic systems of the substituents on the Si–Si bond. The Si–Si bond is a chromophor. With increasing chain length the UV maxima shift to longer wavelengths. In polymeric forms, the absorption maxima reach the visible region and the compounds become yellow (see Chapter 10: Polymeric Compounds). An electron delocalization seems to exist between the Si–Si bonds.

The connection between the bathochromic shift chain length is especially clear in derivatives with substituents which have little or no influence on the Si–Si bond (such as the methyl group). Neither does the acyl group appear to have any influence on the Si–Si bond (since the UV spectra of acylsilanes and acyldisilanes are similar[55]), nor does the Si–N bond system[454].

Investigations of vinyl derivatives, however, show a stronger interaction between the olefinic bond system and the Si–Si bond[87, 181, 499]. This interaction was also theoretically substantiated by an antisymmetrized free-electron molecular-orbital method[15], and Hückel molecular orbital (HMO) calculations confirm these results[457].

The influence of different groups on the electronic absorption of compounds of the type: $me_3Si–Sime_2–Sime_2X$ and $(Sime_3)_2SimeX$ with X = H, halogen, O-, N- and S-groups was measured[450]. It was found that the long-wavelength absorption band appears to originate from common transitions and there is thus no reason for specifically implicating d orbitals.

The interpreation of UV spectra becomes difficult when effects of σ delocalization of the Si–Si bond are mixed with d_π interactions from the substituents. The strongest bathochromic shift occurs with phenylated polysilanes when the first phenyl group is substituted.

Recent studies in photoelectron spectroscopy show good agreement with electronic absorption spectra[451] and show very little or no influence of the d orbitals. The good correlation between the vertical ionization energy and the HMO eigenwert coefficients with methylated chain, branched chain, and cyclic polysilanes shows the applicability of the Koopman theorem. The influence of the antibonding molecular orbitals and a π-bond interaction with the unoccupied Si orbitals is negligibly small[42].

In phenyl-substituted polysilanes, the problems become more complicated because the benzene π electron system is mixed with the σ (Si–Si) orbitals and the highest occupied molecular orbital is assigned to a linear combination of π (C_6H_5) and σ (Si–Si) orbitals[452]. Similar properties are found in mixed-substituted methyl-phenylpolysilanes, where the effect of extension of the 3 d_π molecular orbitals of the polysilane-chain is found to be minimal[453].

7. Silylenes

Silicon in the lower oxidation states is discussed by Bürger[67a]. Silicon compounds in the divalent state also exist with formation of Si–Si bonds. In the carbon group the tendency to divalent state increases with higher atomic numbers while the tendency to tetravalence declines. Since silicon atoms are relatively small, the tetra valent state is strictly preferred.

The most likely compounds with a silicon atom in the divalent state would be those with strongly electronegative elements. Such substituents would have a strong electron-withdrawing effect for the non-bonding electrons of the silicon atom, producing stabilization. The most theoretically feasible compounds are SiF_2, SiO and $SiCl_2$, but the stability of such compounds is not expected to be very great. However they do exist, and in the following pages a review of these compounds and their properties is given[18, 433].

First, some remarks about nomenclature: Various names are used for compounds like SiX_2, one of the most logical being „silylene", which we use here. It shows the analogy with the carbon-analogs CX_2 which are derivatives of methylene. Derivatives of the silylene SiH_2 are called, for example difluorosilylene SiF_2, dimethylsilylene $Sime_2$, monomeric silicon monoxide would consequently be named oxysilylene.

SiH_2 was first observed as a metastable intermediate of the thermal or photolytic cleavage of silanes[115]. First observations made by Stokland[606] indicated the probable existence of SiH_2. Detailed investigations by Purnell and Walsh[460] show the existence of SiH_2 as a short-lived intermediate of the reaction of silane with phosphorous hydride, PH_3, by neutron irradiation[162], which yields silylphosphine, $H_3Si–PH_2$.

The pyrolysis of disilane forms mono-, tri-, and n- and i-tetrasilane. An insertion reaction takes place[51] in which silylene reacts with an Si–H bond:

$$Si_2H_6 \longrightarrow SiH_4 + :SiH_2$$
$$H_3Si\text{–}SiH_3 + :SiH_2 \longrightarrow Si_3H_8$$
$$\text{etc.}$$

A similar insertion reaction is possible with an Si–Si and an Si–C bond, which explains the formation of $CH_3SiD_2SiH_3$ from the reaction of Si_2D_6 with methylsilane. Because it is easy to form silylene by a pyrolytic decomposition of silane, other similar reactions can be expected to be discovered in the future.

It is of interest that carbene, CH_2, generated from CH_2N_2 does not give an insertion reaction with the Si–Si bond in Si_2me_6 but reacts with the Si–C linkage, forming ethylpentamethyldisilane[93)].

The results of spectroscopic investigations of SiH_2 indicate a singlet state of SiH_2, the molecular parameters being calculated[115)] to be r SiH = 1.521 Å and the angle HSiH = 92.1°.

One of the most thoroughly investigated silylenes is difluorosilylene. The oldest observation was made by Troost and Hautefeuille[639)] in 1871, who postulated the existence of SiF_2 in the transportation reaction of silicon with SiF_4:

$$Si + SiF_4 \rightleftharpoons 2\ SiF_2 .$$

From thermodynamic investigations the heat of formation in the above reaction was estimated to be −148 kcal/mole[409)] and −139 kcal/mole[329)]. The differences are explained by the existence of different polymeric particles formed at different temperatures in the vapor phase. Recent work by Margrave, Timms and coworkers[633, 634)] showed the existence of SiF_2 in preparative yields (up to 90%) with the same reaction at 1150 °C. Not only silicon metal but also silicon alloys, SiC, silicides etc. can be used for this reaction. SiF_2 is formed as a gas which condences at −196 °C to a reddish-yellow compound. It becomes colorless and rubberlike upon warming to room temperature. It burns in moist air and evolves hydrogen with aqueous alkalis. With an aqueous solution of hydrogen fluoride, silanes like Si_nH_{2n+2} are formed[633)] (n = 1–6). Heating in vacuo causes decomposition and the formation of silicon fluorides with low molecular weights:

$$(SiF_2)_x \longrightarrow Si_n\, F_{2n+2} + 2\,(SiF)_{polym}.$$

with n = 1 up to 14.

The rubberlike colorless form of SiF_2 is a polymeric molecule.

Monomeric SiF_2 has a half-life (approx 150 sec at 0.1 mm Hg) 100 times longer than the carbenes. It is stable enough to be available for other investigations and reactions.

Other methods for the preparation of SiF_2 are known. It is formed by electrical discharge in SiF_4[297, 464)]. Pyrolysis of Si_2F_6 yields monomeric SiF_2[594)], which polymerized very quickly under the given reaction conditions. It formed a colored precipitate which become colorless on warming to room temperature.

Detailed spectroscopic investigations were done with monomeric SiF_2. IR[25, 380)], UV[379)], ESR[298)] and microwave[462, 463)] spectra have been measured, the results indicating nonlinear molecule with a bond angle of 100° 59′ and interatomic distance of 1.591 Å. (Si–F). The observed bond angle relates to other observations of carbene-like molecules and shows the carbenoid character of the difluorosilylene. The interatomic distance is longer than in tetrafluorosilane, in which it is 1.54 Å. This also relates to a postulated singlet state in which there is more *p* character in the Si–F bond system.

The IR spectra, measured in a gaseous SiF_2/SiF_4 mixture[380] and in a noble gas matrix at low temperature[25], show the expected wavelengths, a symmetric stretching-frequency at 855 cm^{-1} and an asymmetric stretching frequency at 872 cm^{-1}. A bending mode was observed at 345 cm^{-1} [379]. A calculation of the valence force constant gave the value 5.019 mdyn/Å.

The mixture of SiF_2 and SiF_4 obtained in the gaseous state shows no ESR signal because the concentration of the radical is too low, but the yellow-reddish solid gives an unstructured broad signal with a *g* factor of 2.003 ± 0.002[298]. This value agrees well with that expected for a free electron. With increasing temperature, the intensity of the signal decreases until it disappears completely.

The interpretation of these facts is as follows: the monomeric SiF_2 units are first formed, then they undergo polymerization to small oligomers at low temperature:

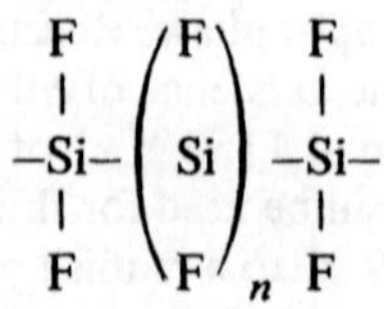

$n = 0, 1, 2 \ldots$

These molecules have radical character, being diradicals in the triplet state. Upon warming to room temperature, the polymerization proceeds until high-polymer compounds without radical properties are formed.

The chemical properties agree with the observed physical properties. Monomeric SiF_2 in the singlet state is not very reactive[629], in contrast to the highly reactive copolymers. Therefore, products with Si–Si bonds often result. As in all radical reactions, competition occurs between polymerization and capture reactions with other reagents.

SiF_2 usually reacts in the form of a singlet molecule as a Lewis base, but a reaction as a Lewis-acid is also possible if the molecule is in the triplet state.

Some examples of the reactivity of SiF_2 will be given. BF_3 is a strong Lewis-acid and it reacts with SiF_2 [522, 632] if the gases are cooled together, giving di- and trisilylborofluorides. The following mechanism is proposed:

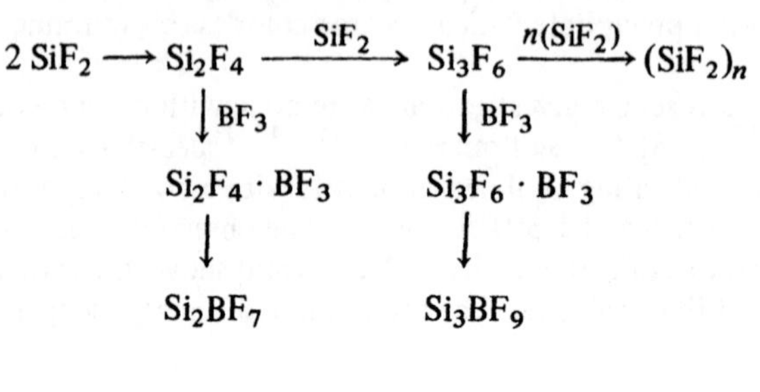

The essential step in these reactions is, apparently, stabilization through an intramolecular rearrangement:

$$F_2Si{-}SiF_2{-}BF_3 \longrightarrow F_3Si{-}SiF_2{-}BF_2 .$$

Reaction with B_2H_6 yields polyfluoropolysilanes like H_3SiSiF_3, $H_2Si(SiF_3)_2$, Si_3F_8 and other similar compounds[524]. With CF_3I, the new compound $CF_3 \cdot SiF_2I$ and some by-products result[411].

IR investigations of the reaction of SiF_2 with NO and CO in a krypton matrix indicate a reaction involving the formation of an addition compound[25]. With PF_3, an unstable polymeric product also results[629].

A typical addition reaction occurs with various Lewis bases, such as bromine (yielding Br_2SiF_2). With water, a formal insertion occurs, forming a silanol[131]:

$$SiF_2 + H_2O \longrightarrow F_2HSiOH.$$

This silanol is not stable and condenses rapidly to a 1,1,2,2-tetrafluorodisiloxane $F_2HSiOSiHF_2$[410]. The first step of the reaction with water yields a colored compound (green or yellow). This colored compound occurs only at low temperatures when SiF_2 and water vapor are condensed together at temperature of −196 °C, indicating that the first step involves the oligomer. The absence of a Si–H frequency in the IR spectrum is consistent with a triplet state in starting material. Hydrogen sulfide reacts with SiF_2 to yield HF_2SiSH, $HF_2Si{-}SiF_2{-}SH$, etc. The reaction should form $HF_2Si{-}SiF_2{-}SH$ as a main product in view of the described diradicals, but the yield of this compound is low. The main product observed is HF_2SiSH. Another mechanism is apparently involved. A similar reaction occurs with germane, GeH_4, in which H_3GeSiF_2H is formed. In this reaction the authors postulate an addition-reaction of GeH_4 to the radical SiF_2 chain in the first step and then a decomposition of the chain[525]. A similar mechanism may occur in the reaction of hydrogen sulfide with SiF_2.

Reactions of SiF_2 with some organosilicon compounds have been recently observed[412]. Tetramethylsilane does not react with SiF_2, and trimethylchlorosilane reacts in only small amounts to yield the corresponding fluorine derivative. Trimethylmethoxysilane yields me_3SiF_3 as main product with $me_3Si{-}Sime_2{-}Sime_2F$ and $meOSiF_3$ as by-products. An insertion reaction apparently takes place with a cleavage of the Si–O–C bond system.

Similar insertion reactions are known with organic compounds. For example, the reaction with methyl fluoride yields fluorinated methylpolysilanes[18]:

$$meF + (n + 1)\, SiF_2 \longrightarrow me{-}(SiF_2)_n{-}SiF_3 . \qquad n = 0{-}3$$

Ethylene reacts to yield cyclic derivatives[631]:

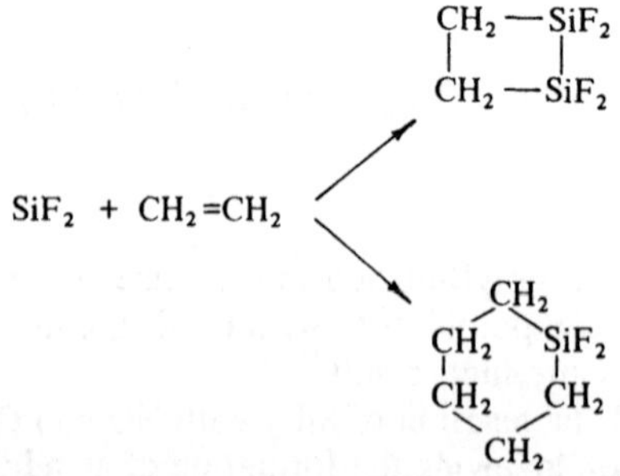

The reaction was carried out with the usual condensation technique, both gases being condensed together at the temperature of liquid nitrogen. Polymers are obtained as by-products.

Similar reactions were investigated with other unsaturated hydrocarbons like butadiene[630], trifluoropropyne[399], allene[400], acetylene[397] and others[398]. Compounds with the SiF_2–SiF_2 group were mainly due to copolymerized SiF_2. Formation of monomeric SiF_2 by the nuclear-recoil technique and the reaction with 1,3-butadiene was shown by Tang, Gennero and Su[626] and yields 1,1-difluoro-silacyclo-3-ene as the major product in contrast to the investigations described above, since the nuclear-recoil technique produces labeled silicon in a concentration far too low to permit dimerization.

The mechanism for reactions of SiF_2 with various fluorine derivatives like trifluoroethylene F_2C=CHF is similar, yielding 1,1-difluoro-2-(trifluorosilyl)-ethane F_2C=$CFSiF_3$, 1,2-difluoro-1-(trifluorosilyl)ethylene FHC=$CFSiF_3$, and 1,2,2-trifluoro-1-(trifluorosilyl)ethylene F_2C=$CFSiF_3$.

An insertion reaction generally takes place with Si–F links in olefinic and aromatic systems forming a C–SiF_3 system. Fluorobenzenes also react in this manner, the mechanism being postulated to involve an intermediate formation of a three-membered ring[637]:

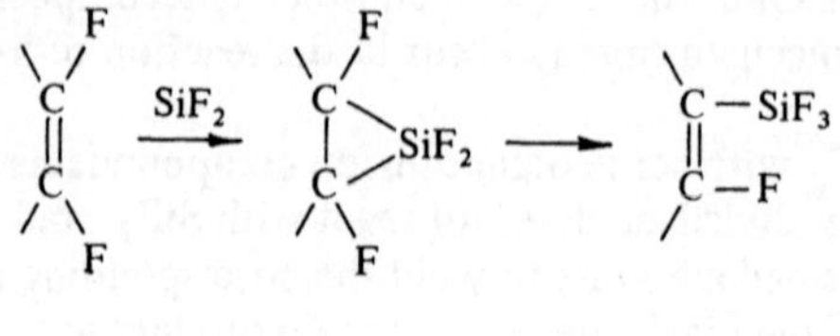

Cyclohexane similarly reacts with SiF_2, but a tricyclic system results[403]:

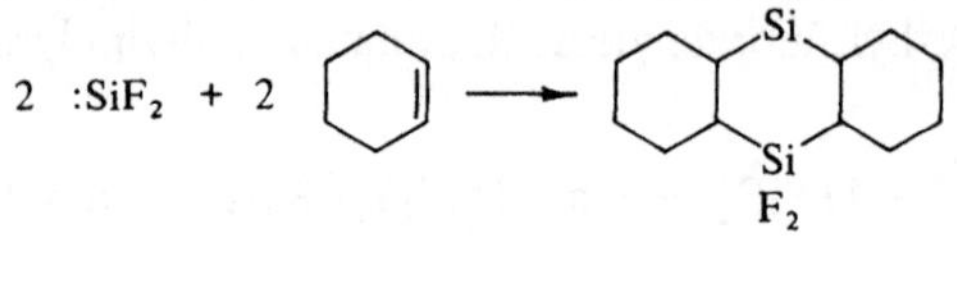

In the reaction with benzene or toluole, the aromatic ring is bridged in 1,4 position with a chain of two or three SiF_2 units:

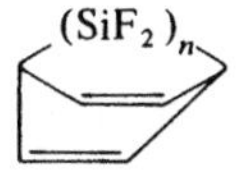

The reaction is probably complicated, with SiF_2 possibly reacting as an oligomer with two or three SiF_2 units when the chain length is suitable for an 1,4-addition.

Other silylenes besides difluorosilylene have been postulated. In many syntheses of polysilanes formed by dimethyldihalosilane with metal, dimethylsilylene is thought to be a shortlived intermediate. We see some aspects of these reactions in the section on cyclic and linear polysilanes.

One can prove the intermediate existence of the silylene by the addition of a radical scavenger, for example, trimethylsilane me_3SiH, which forms pentamethyldisilane[518]:

$$:Sime_2 + me_3SiH \longrightarrow me_3SiSime_2H$$

or with ethylene or other olefins, which form cyclic or linear addition products[433]:

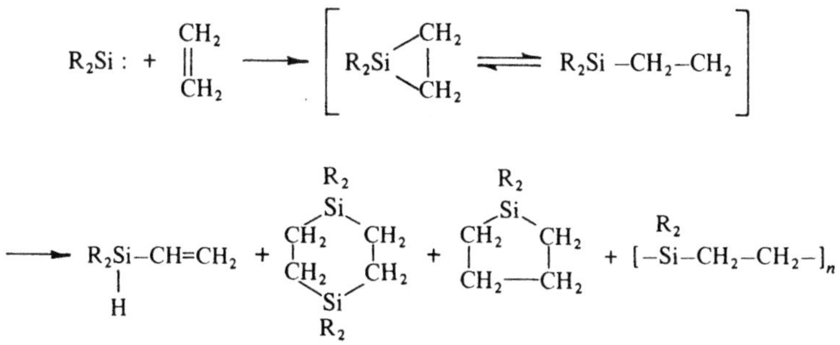

The reaction yields dimethylvinylsilane, but also heterocycles like 1,1-dimethyl-1-silacyclopentasilane or 1,1,4,4-tetramethyl-1,4-disilacyclohexasilane.

The existence and the relatively high stability of the silylene can be explained as a result of the singlet state. If the silylene were in a triplet state, polymerization would proceed faster and the formation of addition products would not be preferred. Herzberg postulated a singlet state, Nefedov[433] found poor reactivity in all electrophilic reactions, and the difference in the reactivity in comparison to organic alkylcarbenes makes the triplet state unlikely.

In the reaction with diphenylacetylene [19, 667] Volpin[649] some years ago postulated a three-membered ring with a carbon-carbon double bond as the reaction product, but West and others showed[321] that the actual product is a six-membered ring and that a three-membered ring is probably an intermediate step:

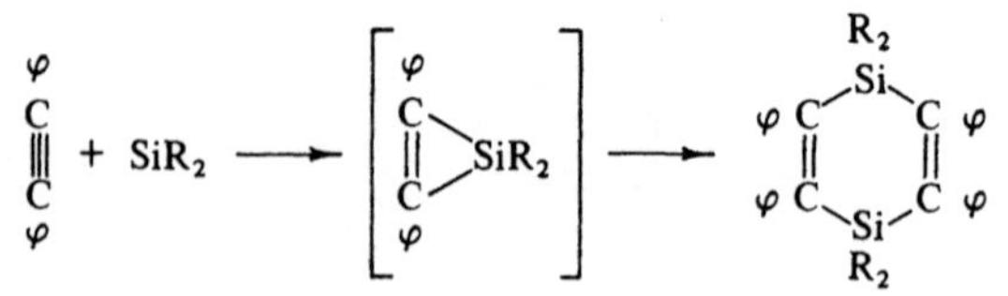

Dimethylsilylene is not only formed in the reaction of dimethyldihalosilane and metals, but also, for example, in the thermal decomposition of metal-silyl compounds[673] like $me_2Al(Sime_3)$ or $Hg(Sime_3)_2$ or in the thermal decomposition of 7-silanobornadiene[188]:

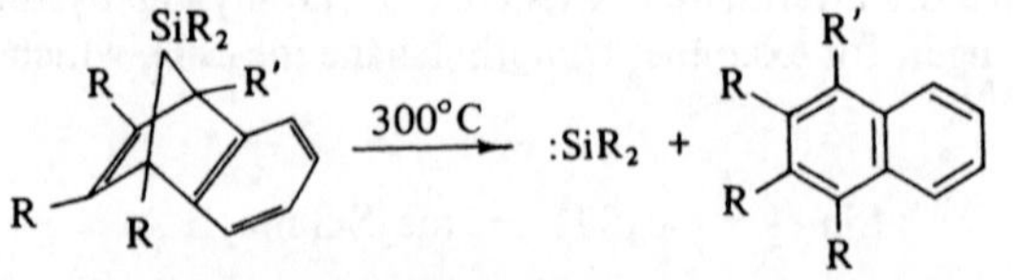

Gilman discovered this reaction after preparing the starting material as follows:

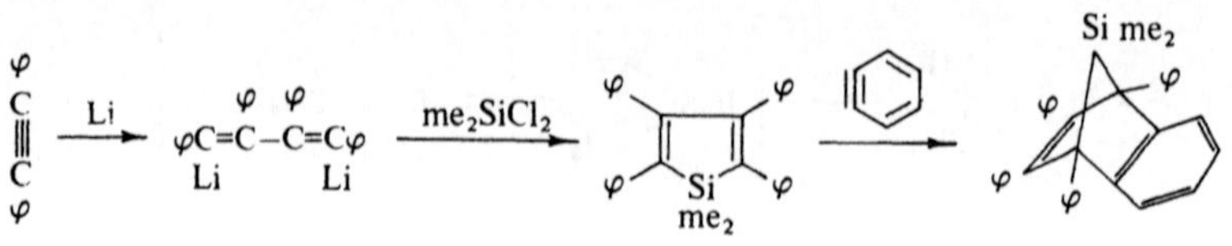

The evidence for the formation of dimethylsilylene from 7-silanoboradiene includes reactions with diphenylacetylene, ethylene, etc., and the observed polymerization to dimethylpolysilane.

7,8-Disilabicyclo[2.2.2]octa-2,5-dienes, prepared by the interaction of the radical anion of anthracene, naphthalene or biphenyl with 1,2-dichlorotetramethyldisilane were found to undergo thermal decomposition[472].

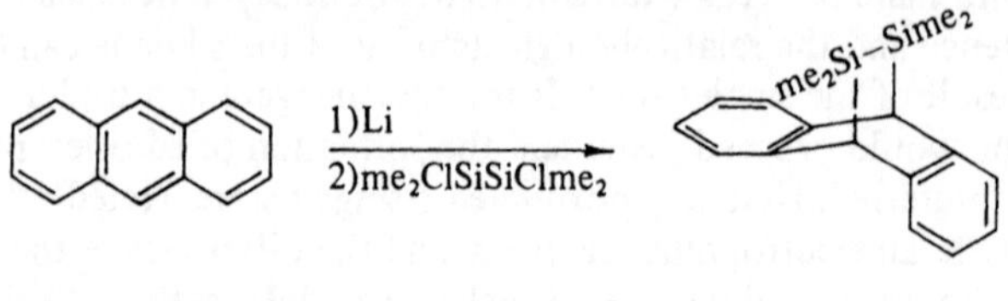

The authors postulate an intermediate containing a silicon-silicon double bond. Stable products include various disilacyclobutanes. A possible mechanism of formation from the postulated $me_2Si{=}Sime_2$ is suggested by the authors.

Polysilane yields monomeric dimethylsilane on heating, whether the dimethylsilane is monomeric or, like the other silylenes (SiF_2 or SiO), a biradical copolymer, is not yet known and more investigations are necessary.

Some new aspects in the chemistry of dimethyldisilanes recently reported by Atwell and Weyenburg include the pyrolysis of dimethyldimethoxysilanes[18–20] to polysilanes and tetramethyldimethoxydisilane:

$$(CH_3O)me_2SiSime_2(Ome) \xrightarrow{225\ ^\circ C} (meO)_2Sime_2 + (Sime_2)_n.$$

This reaction was investigated by kinetic measurements and with addition reactions. These investigations suggest the following mechanism:

$$(meO)me_2Si + Sime_2(Ome) \longrightarrow (meO)_2Sime_2 + :Sime_2.$$

The silylene is formed in the first step and then undergoes an insertion reaction to yield polysilanes:

$$:Sime_2 + (meO)me_2SiSime_2(Ome) \longrightarrow (meO)me_2Si(Sime_2)Sime_2(Ome)$$

or reacts with radical scavengers.

The primary step probably starts with a nucleophilic interaction of the methoxy group with the neighboring silicon resulting in α elimination[17]:

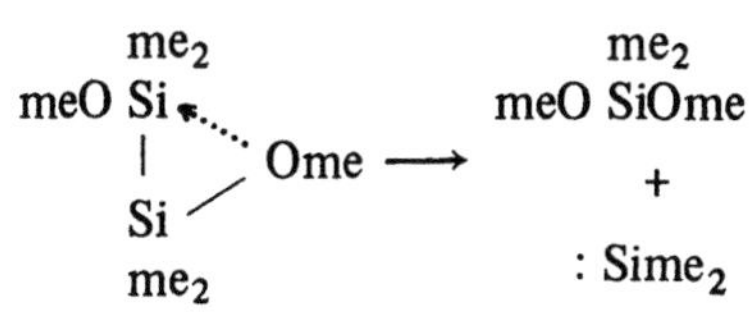

A similar mechanism was postulated in other reactions, for example in alkoxypolysilanes, which react to form alkylpolysiloxanes[553]:

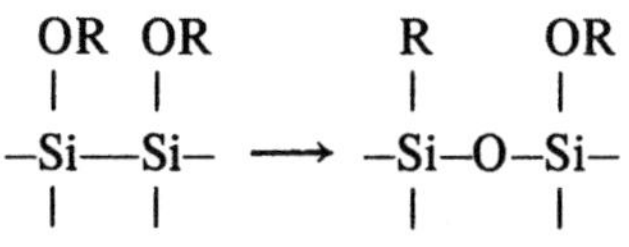

Other methoxypolysilanes like hexamethoxysiloxene[351] react in the same manner, so, it appears that this kind of rearrangement is general in silicon chemistry.

Mixed substituted silylenes occur, such as Sime(Ome), SiHBr, SiHCl, SiHme.

Photolysis of dodecamethylcyclohexasilane[313] and permethylated linear polysilanes[315] yields dimethylsilylene by homolytic cleavage of the Si–Si bonds. Dimethylsilylene readily inserts into Si–H or Si–Ome bonds of appropriate monosilane derivatives. From the photolysis of linear polysilanes it has been shown that the extent of homolysis increases with increasing chain length of the polysilane, whereas the effective yield of the silylene species is virtually constant regardless of chain length.

Inorganic substituted silylenes are formed in the reaction of silicon with silicon chloride[527, 540, 543]:

$$Si + SiCl_4 \rightleftharpoons 2\ SiCl_2$$

the same reaction occurs with $SiBr_4$ and SiI_4[688, 689] (see pp. 105, 106).

8. Silicon Monoxide

Silicon monoxide is one of the most controversial compounds in the chemistry of silicon of lower valence. In 1907, SiO was postulated by Potter[459]) to exist in the gaseous state after which quite a several flood of papers [130,228,569,665,696]) described this compound in both the gaseous and the solid state. The interest in this compound is understandable, because SiO has become technologically important, being a nonconductor, vaporizable in high vacuum and precipitable on surfaces as a glasslike film amorphous to X-rays. SiO and SiO_2 are used as a protective film for lenses and mirrors in the optical industry, on plastic surfaces etc. SiO is one of the best contrast compounds for electron microscopy. In the electronic industry, it is used for gettering in the high vacuum technique and many other applications.

SiO is always present in industrial processes where SiO_2 or silicates are in contact with reducing agents at high temperatures, or where silicon metal, or silicides react with oxidizing agents such as atmospheric oxygen. In both cases, vaporization takes place and yellow or brown SiO condenses on the colder parts of the system.

Only the chemistry and structures will be reported here, since applications are too heterogeneous and too great in number for a complete description in this review.

Results of work on SiO are sometimes confusing and contradictory. One must differentiate between the gaseous and solid states, as each involves a completely different structure.

For the preparation, a one-to-one mixture of SiO_2 and Si is heated:

$$1/2\ SiO_2 + 1/2\ Si \rightleftharpoons SiO.$$

In this reaction, SiO is formed as a gas and then condenses on the colder parts of the apparatus in form of a brown powder. The properties of the condensate are strongly dependent on the condensation conditions, but the type of SiO_2 used is not important[128]).

Other reducing agents (hydrogen or carbon, etc.) give the same results[235]).

The oxidation of silicon to SiO is not dependent on the kind of the oxidizing agent used, and the reaction may be carried out with metal oxides, atmospheric oxygen, etc. The finer the grain of the silicon, the better the reaction.

In all cases SiO is initially formed in the gaseous state. Therefore, the extent of vaporization of SiO_2 or silicates in contact with reducing agents is so large that after the reaction such compounds include no silicon. The existence of the gaseous state also makes it understandable that the silicon source and the type of reaction are not important. The conditions under which SiO precipitates are, however, very important in determining the properties of the solid SiO. A recent preparation method is given by Emons and Helmhold[129].

Fast precipation generally gives a light brown, fine-grained powder, while slow precipitation gives a dark brown, glasslike, brittle product. Some authors have found two products at the same time, for example Emons and coworkers found both a yellow and a brown compound in different parts of the apparatus used.

The already complicated picture of solid silicon monoxide becomes even more so because the analytical composition sometimes differs from what would be expected. Excesses of oxygen are often observed. These divergences are possibly due to a small excess of oxygen in the gaseous phase. It has also been proposed that reactions occur with the surface of the apparatus.

The best way to deal with such problems is first to determine the structure of the gaseous SiO and then to study possible structures for the solid product. *The gaseous state:* The abnormally high vapor densities of a mixture of SiO_2 and Si compared with the single compounds indicates the existence of a gaseous compound. The observed vapor condenses to a product with the Si : O composition ratio equal to 1 : 1. Spectroscopic investigations by Bonheffer[45] and and Saper[508] also support the existence of gaseous SiO. Investigations of the vapor density have been made by various authors[131,541]. The results of Schaefer and Hoernle[541], which are in relatively good agreement with the results of other authors are presented in the diagram (Fig. 1).

The calculated heat of formation varies between –21 and –29 kcal/mole the most recent value being –21.4 kcal/mole[131]. The enthalpy of the equation

$$1/2\ SiO_2 + 1/2\ Si \rightleftharpoons SiO$$

is found to be 87.3 kcal/mole. The dissociation energy

$$SiO \rightleftharpoons Si + O$$

Fig. 1 Vapor density at various temperatures[228]

is estimated to be 170 kcal/mole from thermodynamic data[228)] and 165 kcal/mole from spectroscopic measurements. The dipole moment has been calculated both in the ground state and in the excited state. In the ground state the value is 3.09 Debye[465)].

The results, combined with self-consistent-field calculations by McLean and Yoshimine[407)] lead to the conclusion that there is nearly no modification of the d orbitals of the silicon atom in monomeric SiO and so the bond system of SiO is very similar to that of carbonmonoxide.

Recent investigations of the IR spectrum have used a matrix technique at low temperatures[8,247)] (20 °K in a matrix of nitrogen or argon), enabling investigation of the monomeric form. The results show monomer units of SiO and also oligomers. At temperatures a few degrees Kelvin higher the amount of oligomers increases. Two polymer species predominate, the dimeric and trimeric units, Si_2O_2 and Si_3O_3. Vibrational spectra gave the following parameters:

Table 3: Molecule parameters of Si_2O_2 and Si_3O_3

	Si_2O_2	Si_3O_3
Bond length SiO[473]	1,71 A	1,70 A
Bond force constants	3,55 mdyn/Å	3,73 mdyn/Å
Bond order	1,27	1,29
SiOSi angle	93°	140°
OSiO angle	87°	100°

The structures are rings in which each silicon appears to have biradical character. ESR measurements have not been made yet. The possibility that a larger ring like $(SiO)_5$ may occur has also been proposed.

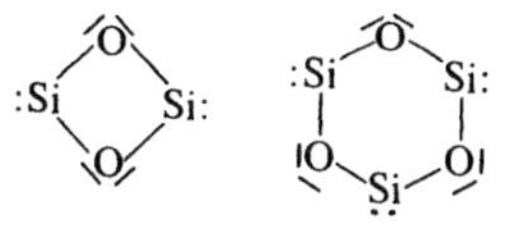

Two important conclusions can be drawn from these investigations. Firstly, these oligomers are probably the first step in the polymerization which precedes condensations to the solid state. Secondly, there is a great similarity to the SiF_2 system; similar oligomers $(SiF_2)_n$ with $n = 2$ and 3 are formed from monomeric SiF_2.

SiO may thus react in similar ways to SiF_2 (for reactions with organic compounds see p. 72).

SiO in polymeric state: There is good knowledge of SiO in the gaseous state, but not in the solid state. SiO is thermodynamically unstable at room temperature. It disproportionates to silicon metal and SiO_2. Under some reaction conditions the product sometimes shows the X-ray lines of silicon. The silicon grains are often so small that the compound appears uniform. Some authors consider that pure SiO does not exist in the solid state.

In view of the instability at room temperature, Hoch and Johnston[288] investigated SiO in the solid phase at 1300 °C and found by X-ray analysis a new lattice, a cubic cell with a = 7.135 Å, which they postulated to be the SiO lattice. Decomposition occurs when the product is cooled slowly and the X-ray lines of silicon and SiO_2 may be seen below 850 °C. With rapid cooling, *e.g.* from 1300 °C to 850 °C over 2 minutes, the SiO lattice is frozen as a metastable system.

This view is opposed by Geller and Thurmond[169], who think the lattice described above is a compound with the carbon from the binding agent used in the X-ray analysis. It has so far remained unresolved, which, if either, of these explanation is correct.

Independent of this dispute concerning the crystalline state, the existence of amorphous SiO in the solid state may also be possible. Electron diffraction investigations show specific lines in some preparations. Emons found that only the dark brown product shows these lines. The yellow product shows only the lines of silicon and silicon dioxide. These observations suggest the possibility of the existence of a solid, probably amorphous SiO.

Some work on X-ray fluorescence has been done by Cremer[97] and others. The wavelength of a K_α line of a given element is normally characteristic for the element, though small differences determined by its oxidation state are observed (Table 4).

Recent investigations by Faessler[133], mainly on sulphur and phosphorus, show that the X-ray absorption shifts to a few Angstroms shorter wavelengths indicating higher energies when the oxidation state of the element is raised.

Few measurements have been done with silicon in general and only one or two investigations were done with silicon monoxide. The results of the SiO measurements gave lines which correspond to an oxidation state of two, but other products indicate the existence of an oxidation state of three[97].

Table 4: K_α-lines of various silicon oxides

	$K\alpha_1$	$K\alpha_2$	$\delta\lambda$ mean	δE
Si	7111.16	7113.77	–	–
Si_2O (in the form of SiOR)	09.32	12.00	1.81	0.44
Si_2O_3	08.89	11.55	2.25	0.55
SiO_2	08.69	11.36	2.44	0.60

So far, these results have been too limited to permit exact conclusions. However, silicon would be expected to appear in different oxidation states, in view of at the following structures:

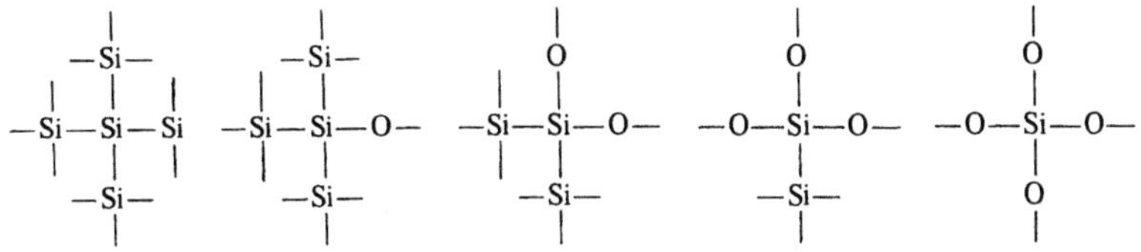

0 +I +II +III +IV

An amorphous silicon monoxide could be formed as a structure between metallic silicon (oxidation state zero) and silicon dioxide (oxidation state four). Structures like those shown above seem to be possible. If we summarize all available facts about SiO in the solid state, the following picture is obtained:

Monomeric gaseous silicon monoxide is formed like carbon monoxide, then polymerizes to four – or six-membered rings. These cyclosilanes are radical and unstable. During the precipitation, these radicals undergo a further polymerization. This polymerization forms regular or irregular structures, depending on the conditions of the precipitation. A regular cubic structure may be formed at temperatures over 1300 °C, and at lower temperatures, an irregular structure may occur. The copolymers Si_2O_2 and Si_3O_3 react with one another and form structures with Si-Si bonds.

Such structures are not stable and undergo a disproportionation depending on the precipitation conditions. With very fast cooling, a metastable system seems to undergo a disproportionation, while slower cooling gives silicon metal and silicon dioxide. This disproportionation takes place either completely or only partial, depending upon the precipitation conditions. The grains formed can be small or large as determined by X-ray measurements, even without recrystallization.

The SiO system is thus not simple, and its complexity increases with changes in stoichiometric composition. Oxygen in a small excess in the vapor phase results in slight oxidation. If reducing agents are present, for example if the surface of the apparatus has reducing properties, a small amount of silicon metal is produced.

Summarizing, we can say that SiO in the solid state is generally a mixture of different atomic arrangements and bond systems produced in different preparation conditions. Our knowledge is not extensive enough to encompass all details, but the outline is coherent.

In accordance with such a concept are the results of Timms and Phillips[635)] who observed formation of higher polysilanes Si_nH_{2n+2} (with n = 1,2,3,4,5) in the reaction of SiO with HF. The Si–O bond is cleaved by HF, whereas the Si–Si bond is not. The formation of higher silanes thus shows the presence of small Si-Si units with a small and varying number of Si-Si bonds.

Finally, some recent developments in this field will be mentioned. The formation of the oligomers Si_2O_2 and Si_3O_3 shows a strong similarity to the SiF_2 system. The reactivity is extremely high, due to the radical character. Therefore a reaction with organic material would be expected to form organosilicon compounds. A recent first attempt by Schaschel, Gray and Timms[544)] to condense gaseous SiO together with the organic material at −196 °C, gave the expected reaction. The experimental technique is the same as that used in the SiF_2 reactions.

Hydrocarbons, alkenes and alkynes react with SiO to form white solids, which are stable up to 350 °C. Their structure is unknown: the IR-spectrum indicates Si–O, C–H and Si–H bonds. The SiO probably reacts with the C–C double bond and also with the C–H bond:

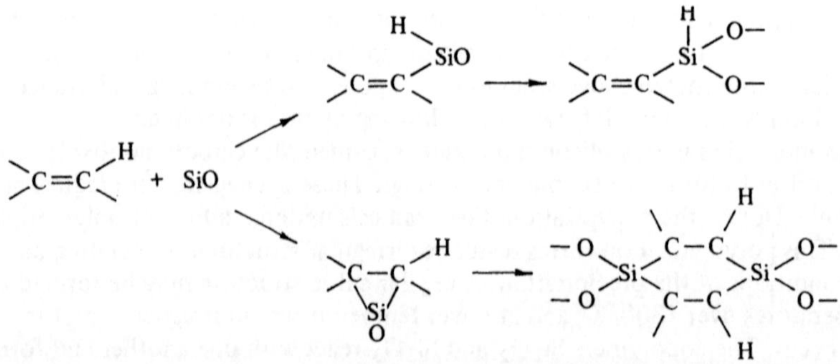

Polymerization to six-membered rings or more complicated systems follows. Hydrolysis yields the starting alkene, the corresponding alkane, and some hydrogen.

There is also the possibility of a similar reaction between the copolymers Si_2O_2 and Si_3O_3, though reaction mechanisms are so far unknown.

Acetylene reacts with SiO to yield a pyrophoric white compound and diphenylacetylene yields a yellow solid, stable in air up to 400 °C. The structures of these compounds are also unknown.

A 1,4 addition seems to occur with dienes like 1,3-butadiene derivatives and benzene. The formation of a yellow compound by reaction of SiO with benzene results in a compound stable in air up to 500 °C. The appearance of Si-H vibrations in the IR spectrum suggests an insertion reaction in the C–H bond, similar to the reaction of SiF_2 with benzene.

Very stable products with compounds containing Si–H bonds are also formed with alkanes. Hydrolysis yields the starting alkane:

$$\gt C{-}H + SiO \longrightarrow -C{-}\overset{O}{Si}{-}H \xrightarrow{H_2O} \gt C{-}H + HO{-}\overset{O}{Si}{-}H.$$

Halogenated hydrocarbons undergo an insertion reaction in the carbon-halogen bond and, in the case of fluorine, the reaction products are explosive, probably because of the large energy of formation of the Si–F bond.

Solid products are also formed with water, ammonia, alcohols and amines, when an insertion reaction probably also occurs.

The recent realization that SiO reacts like SiF_2 and that SiO is a silylene has not yet been fully investigated. Hence, one can expect some interesting development in the future in this field.

9. Cyclic Polysilanes

Our knowledge of the chemistry of cyclic silanes[12,211] is very limited and incomplete and is roughly equivalent to the state of knowledge of cyclic organic compounds in the middle of last century.

Isocyclic silanes are only known in the form of derivatives. Unsubstituted cyclic hydrosilicons have not been investigated, though Si_5H_{10} was recently prepared (see p. 6). The best known derivatives are those with the simple organic groups like methyl or phenyl. Derivatives with other groups are rare.

Some inorganic cyclic silanes with halogen substituents have been investigated, thoug their properties are almost unknown, and even their existence is controversial.

Isocyclic and some heterocyclic organo-silanes are well known, including four-, five-, six- and seven-membered rings, and an eigth-membered ring is also believed to have been observed. The stability of these compounds is similar to that of their carbon analogs. Five- and six-membered rings in silicon chemistry also show the highest stability and the least ring-strain. The five-membered ring seems especially stable, as described later.

9.1. Isocyclic Silanes

9.1.1. Methylated Isocyclic Silanes

The first compound of this type was found by Burkhard[68] in 1949; he reacted dimethylchlorodisilane with sodium as in a Wurtz synthesis. The main product was a polymeric chain with alow yield of the six-membered ring. Gilman[221] and Hengge[273] were able to increase the yield to 70% by changing the reaction conditions. Stolberg[608] obtained a yield of 80%, starting from 1,2-dichlorotetramethylsilane. Dodecamethylcyclohexasilane, a white crystal, is very stable in air and soluble in most nonpolar organic solvents.

The chemical reactions of this compound are not known very well. One can completely destroy the ring to means of an alkaline hydrolysis, a reaction which yields simple units with one silicon atom. Gilman and Inoue[198] cleaved the ring with phosphorous pentachloride or tetrachloroethane to Si chain compounds of various chain lengths, those from four to six predominating, but including smaller chains

$$me_2SiCl_2 \xrightarrow{Li} Si_6me_{12} \xrightarrow[Cl_2CHCHCl_2]{PCl_5} Cl(Sime_2)_nCl. \qquad n = 6, 5, 4$$

The structures of the compounds were identified by reacting them with Grignard reagents which gave the corresponding methyl derivatives, Si_nme_{2n+2}. Cleavage with HCl / butylchloride yields α, ω-dichloropolymethylpolysilanes with different chainlengths[184] Cleavage with K/Na is also possible[609]. The cleavage products undergo some remarkable ring-closure reactions. Halogenated chains and dilithium polymethylpolysilanes react with alkali metals and trimethylchlorosilane respectively to form rings; the chain length of the reactant is not important here, and the main product is always the six-membered ring. These reactions may permit some rearrangements of organometallic compounds, chiefly the lithium compounds. Formation of the six-membered ring is obviously preferred as it has the least amount of ring strain, though recent investigations demonstrate the existence of other ring sizes. Linear polysilanes, phenylated at the ends, react with catalytic amounts of triphenylsilyl lithium to form various permethylated silicon rings with $Si_2\varphi_6$ as a by-product[361]. The reaction apparently starts with a metal rearrangement:

$$\varphi_3Si(Sime_2)_nSi\varphi_3 + \varphi_3SiLi \longrightarrow Si_2\varphi_6 + Li(Sime_2)_nSi\varphi_3$$

and some cleavages, for example:

$$\varphi_3Si(Sime_2)_nSi\varphi_3 + \varphi_3SiLi \longrightarrow \varphi_3Si(Sime_2)_{n-1}Si\varphi_3 + Li(Sime_2)Si\varphi_3.$$

The Li compounds thus formed are able to react with the starting material, resulting in different chain lengths:

$$\varphi_3Si(Sime_2)_nSi\varphi_3 + Li(Sime_2)_xSi\varphi_3 \longrightarrow$$

$$\sum_{x=0}^{x=6} \varphi_3Si(Sime_2)_xSi\varphi_3 + \sum_{x=0}^{x=6} Li(Sime_2)_ySi\varphi_3.$$

The formation of various cyclic silanes results from an intramolecular φ_3SiLi elimination:

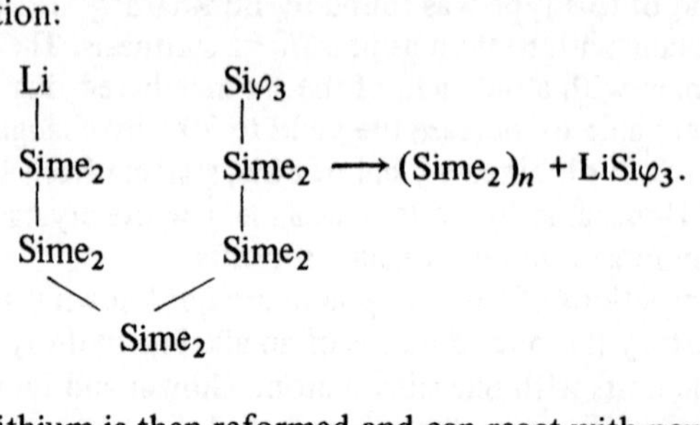

Triphenylsilyl lithium is then reformed and can react with new starting material.

Other methylcyclosilanes of various ring sizes have been observed by West[74, 75]. Under different reaction conditions, the reaction of dimethyldichlorosilane with Na-K alloy yields compounds of different ring sizes. The various compounds were separated by gas chromatography. He found five-, six-, and seven-membered rings, but the existence of four- and eight-membered rings is also probable. West has determined the best conditions for the formation of each ring size. A kinetic preference for the formation of the five- and seven-membered ring was found to exist, but there is a thermodynamic preference for the six-membered ring. Short reaction times generally give mainly five- and seven-membered rings, while longer reaction times give predominantly the six-membered ring.

In a typical run, dimethyldichlorosilane was dripped for one hour on to Na-K alloy to a 10% excess, followed by refluxing for one hour in THF, separation and hydrolysis. The yield cyclic material was 50%, comprising 20% Si_5me_{10}, 8% Si_7me_{14} and 72% Si_6me_{12}.

Bromides or iodides cause complications because cleavage occurs in THF. In nonpolar solvents like cyclohexane, the yield is very low and no rearrangements to other ring sizes are observed. Polar solvents therefore seem necessary for rearrangements.

Thermal decomposition of linear polymethyl-silanes[75,537] and alkalicatalyzed rearrangements of linear polymers give cyclic compounds, chiefly the six-membered ring[516].

Cyclic methylpolysilanes very readily undergo rearrangement reaction. By refluxing pure decamethylcyclopentasilane in THF in the presence of naphthalene radical anions, 10% of the cyclic hexamer is formed after 1 h.

One of the outstanding properties of methylcyclosilanes was found by West[76, 301]. The three known compounds $(Sime_2)_n$ with n : 5, 6, 7 react to give radical anions. This formation may be effected via electrolytic reduction or by reacting the $(Sime_2)_n$ compounds with K/Na alloy. The most stable anion is the five-membered ring, which West attributed to its having a greater degree of planarity than the others. The anion is dark blue and stable only in solution. The ESR investigation shows a complete delocalization of the electron over the whole ring. The dark color and also HMO calculations suggest a π system in the ring[457], Gohlke[229] observed a negative ion $(Si_6me_{12})^-$ in mass spectral investigations. The UV spectra of the neutral rings show absorption at relatively long wavelengths.

Table 5: UV-spectra of methylcyclosilanes[406]

	λ [mμ]	ε
	210	sh
Si_5me_{10}	261	1100
	272	970
	232	5800
Si_6me_{12}	255 (sh)	2000
	217 (sh)	4500
Si_7me_{14}	242	2100

A six-membered ring anion prepared by electrolytic reduction at −120 °C will rearrange spontaneously to a five-membered ring, even at low temperature and in a short period (half-life 5.3 min at −120 °C).

^{1}H-NMR and IR data are given in Table 6 and 7[76].

Table 6: Proton NMR Data for the Permethylcyclopolysilanes

Compound	Chemical shift ppm	c/s	Coupling constants, c/s ^{29}Si-CH_3	^{29}Si–Si-CH_3	^{13}C-H
$Me_{10}Si_5$	−0.135	−8.1	6.3	3.2	121.0
$Me_{12}Si_6$	−0.132	−7.9	6.4	3.4	120.0
$Me_{14}S_7$	−0.125	−7.5	6.3	3.2	120

Table 7: Infrared Spectra (cm^{-1}) of the Permethylcyclopolysilanes

$Me_{10}Si_5$	$Me_{12}Si_6$	$Me_{14}Si_7$	Assignment
2950 (s)	2950 (s)	2950 (s)	C–H antisym stretch
2890 (m)	2885 (m)	2890 (m)	C–H sym stretch
2795 (w)	2790 (w)	2790 (w)	
1400 (m)	1400 (m)	1400 (m)	C–H deformation
1250 (s)	1250 (s)	1250 (s)	Si–Me deformation
	1240 (m)	1240 (m)	Si–Me deformation
845 (m)	815 (m)	845 (m)	Si–Me rock
830 (m)	830 (m)	830 (m)	Si–Me rock
800 (vs)	800 (vs)	795 (vs)	Si–Me rock
735 (s)	735 (m)	735 (m)	
690 (m)	690 (m)	690 (m)	Si–Me
655 (s)	655 (s)	655 (s)	Si–Me
	630 (w)	630 (w)	
400 (w)	383 (m)	362 (w)	Si–Si stretch

The ionization potentials of these rings were measured by Pitt *et al.*[451]. The potential increases with increasing ring size and is generally higher than for the corresponding linear polysilanes.

Other recent investigations indicate the existence of a four-membered ring. It is white, not stable in air, and apparently has a great deal of ring strain. It was observed in small yields after photolysis of the hexamer[309, 313], but the main product was the five-membered ring. Kumada[307] demonstrated a possible rearrangement from the six- to the five-membered ring by heating the six-membered ring with $AlCl_3$:

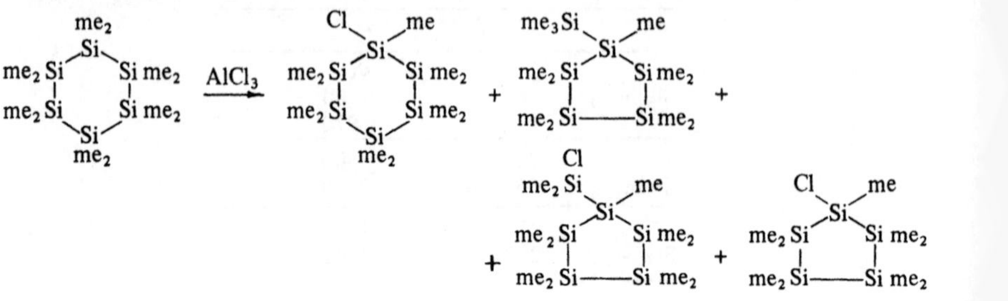

This reaction is the first method discovered for cleaving a Si-methyl bond without cleaving the ring. Another method for breaking the Si-methyl bond was found recently, *i.e.* reacting such systems with $HCl/AlCl_3$[310]. Under the conditions described, this reaction will yield monochloroundecamethylcyclohexasilane, which is able to undergo other reactions involving the reactive Si–Cl linkage:

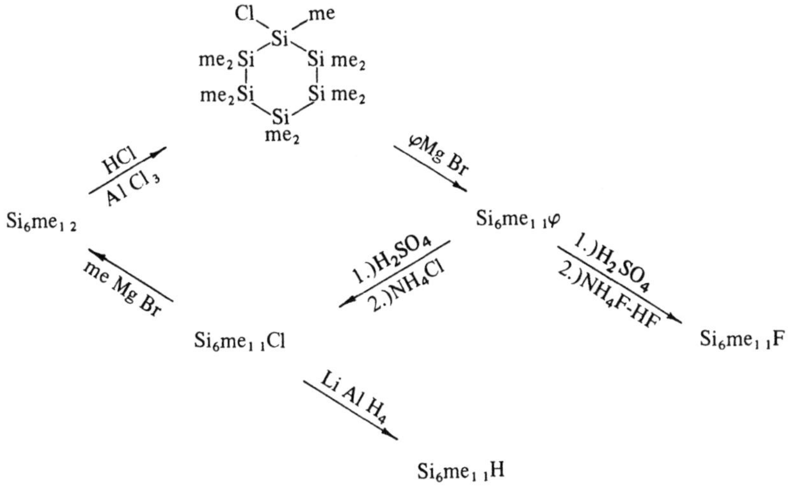

Treatment of Si_6me_{12} with an excess of me_3SiCl in the presence of a catalytic amount of $AlCl_3$ gives chlorodimethylsilylnonamethylcyclopentasilane. Other derivatives are formed by reaction of the chlorodimethylsilyl group with Grignard reagents etc [311].

Finally, a dicyclic system Si_8me_{14} was formed by reacting a mixture of dimethyldichlorosilane with methyltrichlorosilane[303]. This system also gives a radical anion $(Si_8me_{14})^-$ at –90 °C.

West *et al.* recently found[668] that dimethyldichlorosilane and methyltrichlorosilane react with Na/K alloy in the presence of naphthaline in THF to yield bicyclic and cage polysilanes Si_7me_{12}, Si_8me_{14}, Si_9me_{16}, $Si_{10}me_{16}$, $Si_{10}me_{18}$, $Si_{11}me_{18}$ and $Si_{13}me_{22}$. The proposed structures are

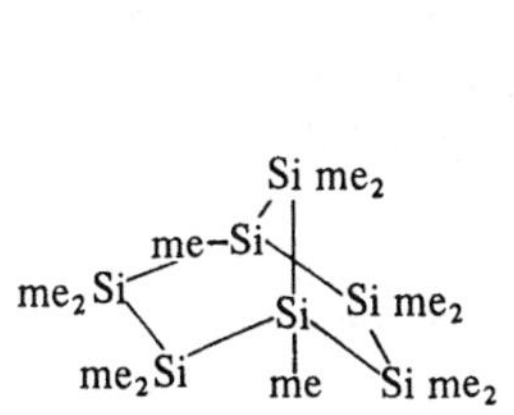

Si_7me_{12}

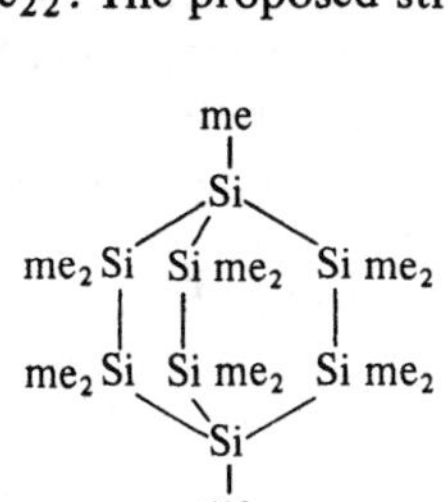

Si_8me_{14}

while the following probable structures are given:

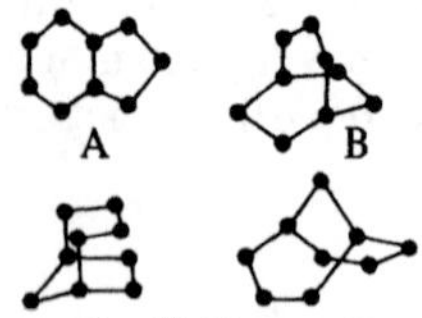

A B

C Si_9Me_{16} D

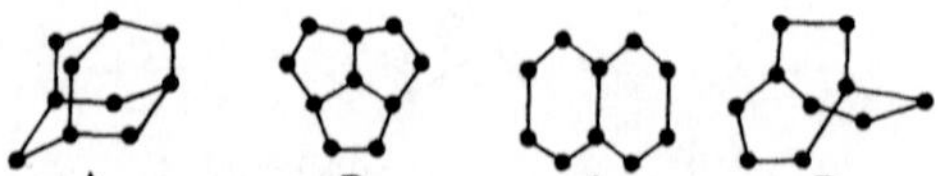

A B A B

3 to 1 $Si_{10}Me_{18}$

$Si_{10}Me_{16}$

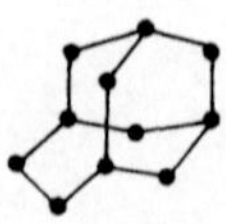

$Si_{11}Me_{18}$

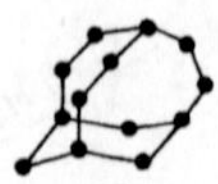

$Si_{13}Me_{22}$

Anion radicals formed by reduction of several of the polycyclopolysilanes were studied by ESR spectroscopy.

From other aliphatically substituted cyclosilanes only a deca(i-butyl)-cyclopentasilane was isolated[302]; preparation of a ring with t-butyl group seems to be prevented by steric hindrance.

9.1.2. Phenylated Cyclosilanes

The oldest known cyclosilanes are the phenyl derivatives. Perphenylated cyclosilanes $(Si\varphi)_n$ with $n = 4,5$ and 6 are known. Their high molecular weight, chemical stability and low solubility led to some mistaken conclusions regarding ring sizes and structures[206, 274, 682], but today the picture seems clearer.

The first work was done by Kipping in 1921[386]. He synthesized some cyclic perphenylated silanes from diphenyldichlorosilane by reaction with sodium and called them A, B, C and D. It would take too long to tell the whole story of past errors, but present opinion is that compound A is octaphenylcyclotetrasilane. This four-membered ring has the most ringstrain and greatest reactivity. The greatest yield (75%) is obtained with lithium in THF[319]. Magnesium catalysis also yields the four-membered ring[207]. The known reactions of this ring are indicated in reaction Scheme I:

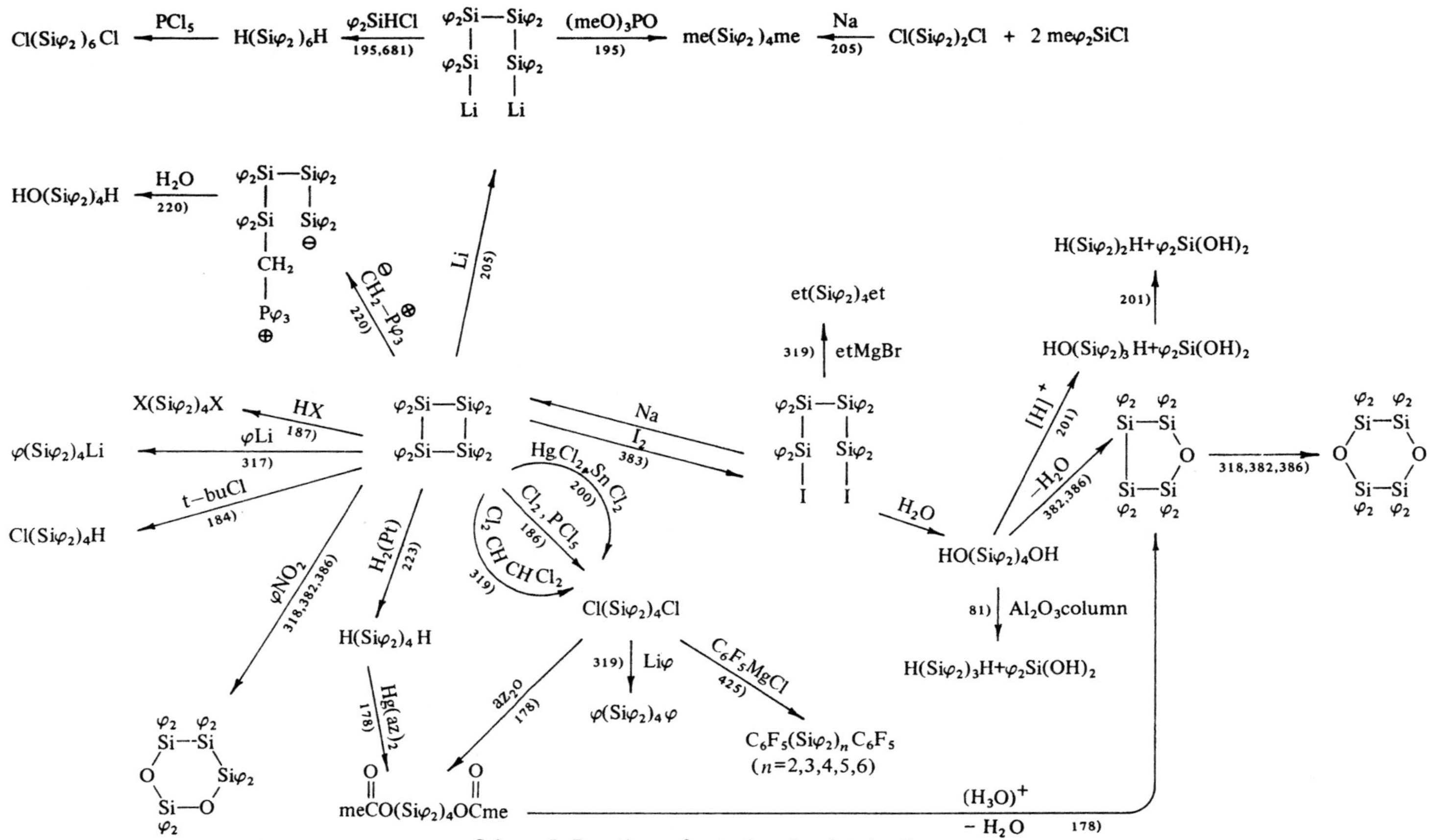

Scheme I: Reactions of octaphenylcyclotetrasilane

Iodine cleaves the ring to a di-iodide[383, 387], which reacts with water to yield the unstable dihydroxy compound[318, 387] that reacts spontaneously to oxygen-containing heterocycles[318, 387]. Two dioxides are also formed. The properties of these compounds are described in the chapter on heterocyclic silanes. Chlorination occurs with mercury chloride[200] and tetrachloroethene[319]. Chlorine cleaves the ring to form the disilane[186]. When the chains are halogenated at the ends, they react with methyl lithium, Grignard reagents or Li phenyl[319] to form the corresponding derivatives.

With lithium, an important cleavage occurs, giving a chain with two Li atoms in the end positions[205]. The di-Li compound is usually formed only in solution. It recently proved possible[280, 281] to isolate the pure compound $Li_2(Si\varphi_2)_4 \cdot 2$ THF. This is important because of the number of reactions that begin with this step. One of simplest is the reaction with trimethylphosphate which yields the corresponding methyl derivatives[195, 205, 319], but heterocycles can also be formed (see p. 87).

With diphenyldichlorosilane, a dodecaphenylhexasilane is formed[195] and the two hydrogen atoms can than be halogenated by PCl_5. More reactions of this dilithium compound are known, but the mechanisms are undetermined. Rearrangements occur when A is reacted with phenyl lithium[317], triphenylsilyl lithium or Grignard reagents.

These organometallic compounds split the four-membered ring and form more stable ring systems, especially the five-membered ring.

During the reaction, smaller chains may be formed since cleavage does not stop after the initial splitting of the ring. The transitory presence of silylenes is also possible[313].

Ring cleavage occurs with hydrogen halides[187], hydrogen[223], and mercury acetate[178], yielding the expected compounds. The cleavage products with methylentriphenylphosphorane after hydrolysis are 1,1,2,2,3,3,4,4-octaphenyltetrasilane-1-ol in a small yield and other unidentified products[220]. The cleavage reaction probably starts with ring cleavage by the CH_2 group:

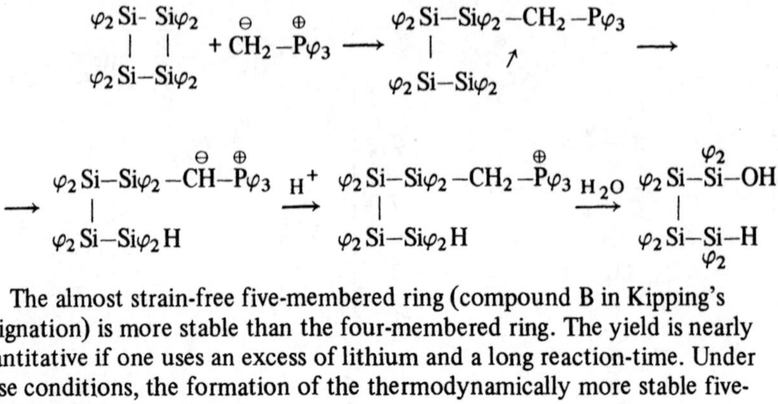

The almost strain-free five-membered ring (compound B in Kipping's designation) is more stable than the four-membered ring. The yield is nearly quantitative if one uses an excess of lithium and a long reaction-time. Under these conditions, the formation of the thermodynamically more stable five-membered ring is preferred.

The reactions of B are similar to those of A, though the cleavage reaction does not take place so easily because it has much less ringstrain. Gilman and Schwebke[209, 210] investigated such reactions, obtaining the results presented in reaction Scheme II:

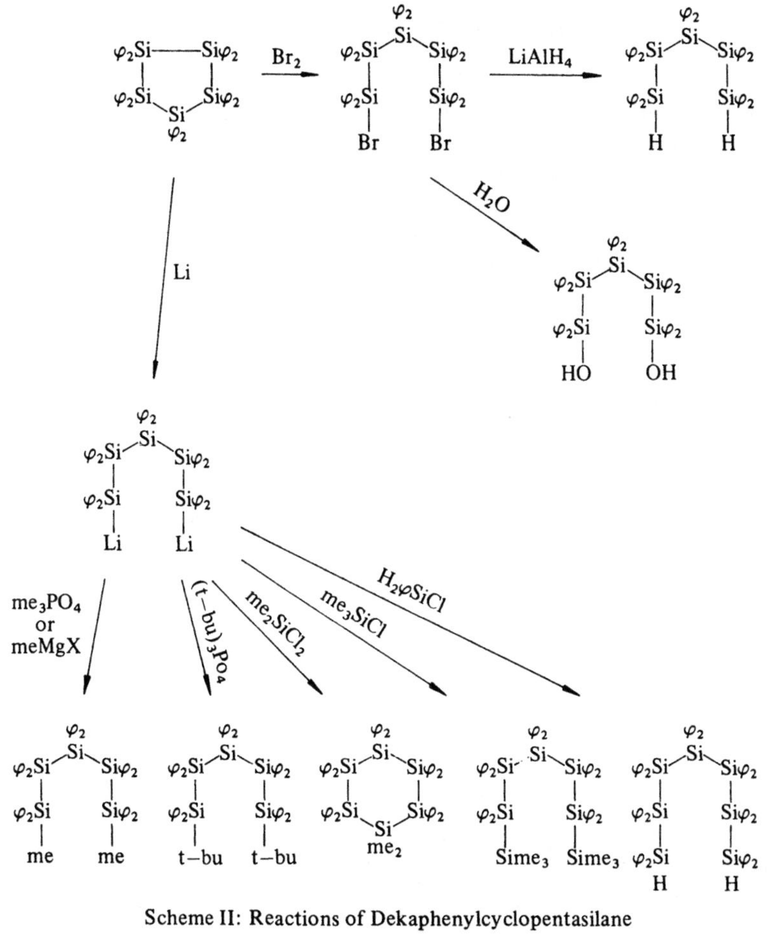

Scheme II: Reactions of Dekaphenylcyclopentasilane

Gilman found that the α, ω-dihydroxy-decaphenylpentasilane formed by a normal cleavage reaction with bromine followed by hydrolysis reacts to form a trisilane when the solution is run through a column of aluminium oxide[81]. The terminal Si atoms are apparently cleaved by the basic OH groups of the Al_2O_3:

$$HO{-}\underset{\varphi_2}{Si}{-}(Si\varphi_2)_3{-}\underset{\varphi_2}{Si}{-}OH \xrightarrow[Al_2O_3]{basic} 2\,(OH)_2Si\varphi_2 + H{-}(Si\varphi_2)_3{-}H.$$

The high stability of the five-membered ring permits a phenyl silicon bond to be split without a cleaving the ring, so, new derivatives of the five-membered ring can be obtained. *E.g.* "A" is split with HI, but "B" is stable to this reagent[157]. Some phenyl groups change place with iodine and benzene is formed:

$$Si_5\varphi_{10} + xHI \longrightarrow Si_5 I_x\varphi_{10-x} + xH\varphi.$$

The reaction proceeds easily until five phenyl groups have gone[264], when it slows down until ultimately 8 to 10 phenyl groups have been removed. With longer reaction times or more drastic conditions, ring cleavage occurs. The first step, the removal of five phenyl groups yields a well-defined product, pentaphenylpentaiodocyclopentasilane, in which each silicon atom has one iodine atom and one phenyl group[265]. This structure is inferred from the cleavage products of the following Grignard reaction:

$$Si_5\varphi_5 I_5 \xrightarrow{CH_3HgX} 5\ me_2 Si\varphi H.$$

The higher iodinated products are probably isomers with various degrees of substitution.

The Si-iodine linkages so formed are very reactive, offering the possibility for forming various derivatives[265]. Methylation to pentamethylpentaphenylpentacyclosilane was effected with methyl lithium. Butyl and methoxy derivatives are also known. Reduction with $LiAlH_4$ gives the hydrogen compound (see reaction Scheme III).

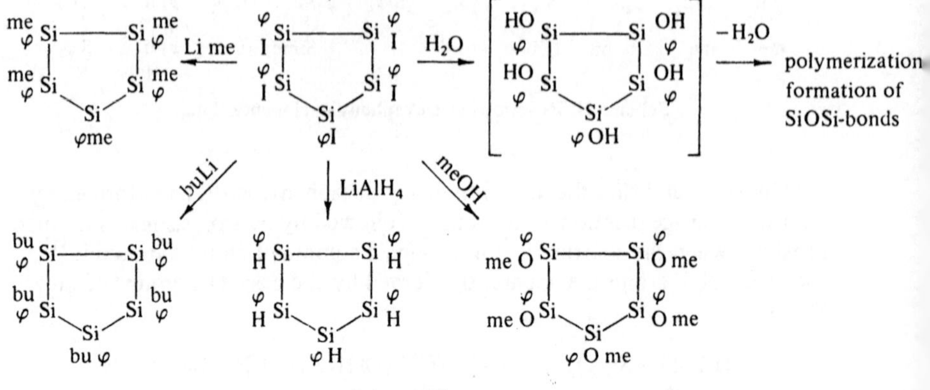

Scheme III

A very interesting reaction occurs with water. At first, the Si–I bond is hydrolyzed and Si–OH bonds are formed. Since silanol groups are not stable, water is eliminated and a condensation polymer results. This compound is yellow and emits a strong yellow Fluorescence. Siloxene, a similarly constructed compound gives the same color effect, which is explained in a later chapter.

Investigations of compound C (Kipping[384]) by Gilman and coworkers[212] indicate that C is probably the six-membered ring. Investigation is very difficult, because C is not very soluble in any organic solvent and its reactivity is very low. No reaction occurs with iodine, basic compounds etc. A molecular-weight determination can only be made by means of mass spectroscopy but this shows only a very small parent-peak[434]. Our Investigations with Raman spectroscopy do not give lines that indicate a six-membered ring, but we have no alternative explanation. More detailed Raman measurements are necessary for a definitive determination of ring-size. Some physical data of the described rings are known: the UV spectra were measured[179] and the IR and UV spectra, X-ray measurements, and thermal decomposition (stable up to 400 °C, the most stable being the six-membered ring) of the described rings were investigated in detail by M'hirsi and Brini[419]. All of the properties of compound D indicate a polymeric character and D appears to be a polymeric chain composed of diphenylsilyl units. Isocyclic compounds with other substituents are rare. Only i-butyl[302] and p-tolyl derivatives[592] are known. Wurtz synthesis was attempted with alkoxydichlorosilanes[273, 388], but a Si–O–C cleavage resulted.

Recent investigations showed the existence of a six-membered ring with two kinds of substituents. Methylphenyldichlorosilane reacts with Li to form hexaphenylhexamethylcyclohexasilane[263]. To prove its structure, the procedure is to remove the phenyl groups of the ring with hydrogen iodide to yield hexaiodohexamethylhexacyclosilane, and then to substitute the iodine atoms with methyl groups by means of methyl lithium. We obtained dodecamethylcyclohexasilane which was identical with an authentic sample isolated by the reaction of dimethyldichlorosilane with Li.

9.2. Heterocyclic Silanes

In the chemistry of silicon, heterocycles are rings containing silicon atoms and one or more of another type of atom. Of the isocyclic compounds, four to seven-membered rings are known, while the existence of an eight-membered ring is probable. The same or similar ring sizes are expected for heterocyclic silicon compounds.

We are familiar with heterocyclic ring systems with various elements, ring sizes, and numbers of hetero atoms. Here only systems with Si-Si bonds are discussed. The number of possible systems is very great in this field with the frequent possibility that different isomers are present. Only a small number of such ring systems have so far been synthesized.

A relatively large number of heterocycles are known in which carbon is the hetero atom. Within this group there are various ring sizes.

A four-membered ring of this type is formed by reacting SiF_2 copolymers with ethylene[631)] or acetylene[397)]:

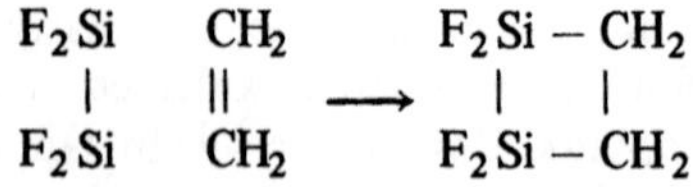

(see p. 62).

A five-membered ring with three carbon atoms and one Si-Si bond was found by some Chinese chemists[654)] who reacted 1-(dimethylchlorosilyl)-3-(methylphenylchlorosilyl) propane with metallic sodium:

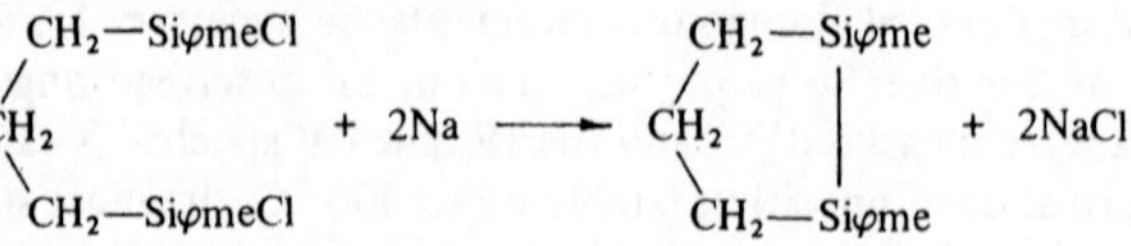

A six-membered ring has also been formed via a rearrangement[493)]:

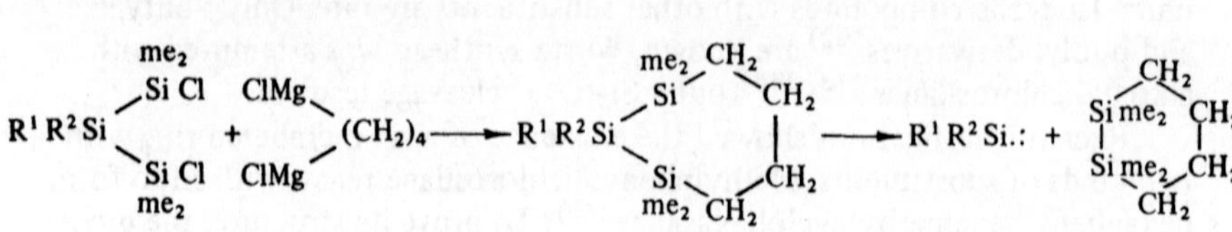

The same ring system was observed by Kumada *et al.* [370, 623)] in considering stereospecificity problems in disilanes (see p. 39).

Another six-membered cycle with two isolated Si-Si linkages was formed by Schmidbaur[554)] while investigating the stability of ylids with disilanyl groups. They found a high tendency to the formation of cyclic ylids:

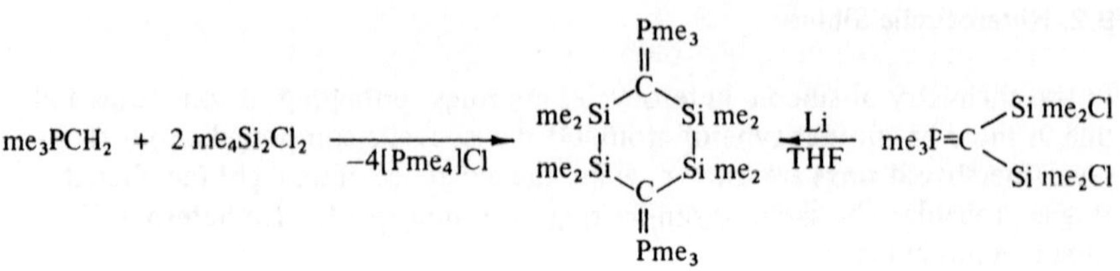

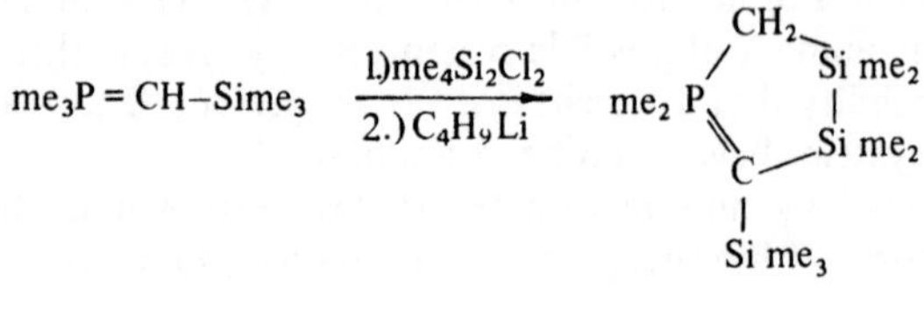

The stabilization of the ylidic centers with Si–Si systems can be explained by a delocalization of the carbanion–charge in the Si–Si system.

Larger ring systems with carbon are formed by a cyclization reaction of dilithium compounds with dichloro compounds[199],

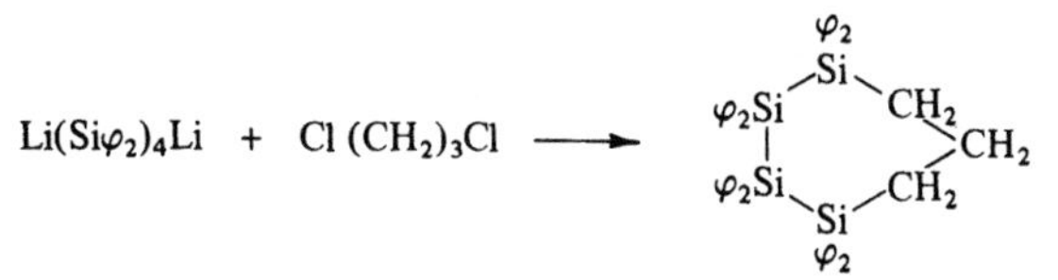

e.g. or by the reaction of dimethyldichlorosilane with lithium in the presence of ethylene. Nefedov *et al.*[432] found high molecular weight compounds with the structure:

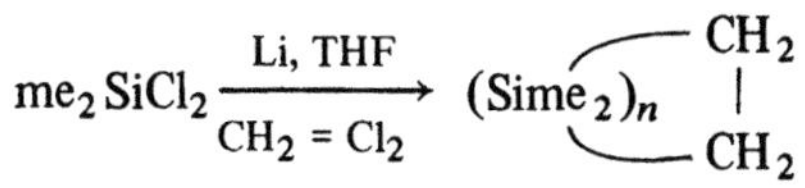

to be main products of this reaction.

Recent investigations by Sakurai *et al.*[493] indicate the existence of some larger carbon-containing rings. 1,1,1,2,2,3,3-hexamethyl-1,2,3-trisilacycloheptane is produced by reacting 1,3-dichlorohexamethyltrisilane with tetramethylene di-Grignard reagent. On irradiation with a low-pressure mercury lamp it gives a six-membered ring, 1,1,2,2-tetramethyl-1,2-disilacyclohexane, and dimethylsilylene. This reaction is a source for silylene radicals (see Chapter 7: Silylenes). An eight-membered ring was also isolated having both methyl and i-propyl groups at position two.

1,1-Dichloro-1-silacyclopentane reacts with phenyldimethylsilyl lithium to yield the disilyl derivative, which can then be halogenated with $HCl/AlCl_3$. The halogen compound reacts with tetramethylene di-Grignard to yield 5,6,11 trisilaspiro-4,6-undecane as an oil.

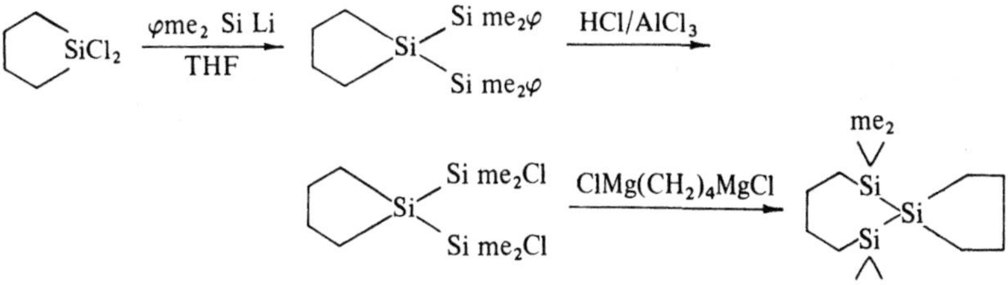

Irradiation of the latter product yields a six-membered ring with one Si-Si linkage, and a heterocyclic silylene

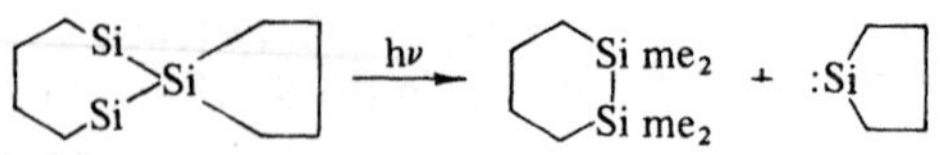

which undergoes a normal insertion reaction into a silicon-hydrogen bond, as does the dimethylsilylene:

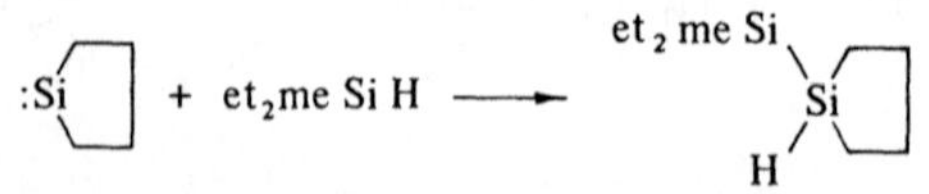

Note that in the investigations of SiF_2 (see Silylenes) a heterocyclic compound was found with carbon as the heteroatoms.

Recent investigations by Sakurai[428)] indicated the existence of a heterocyclic system with C-C double bonds

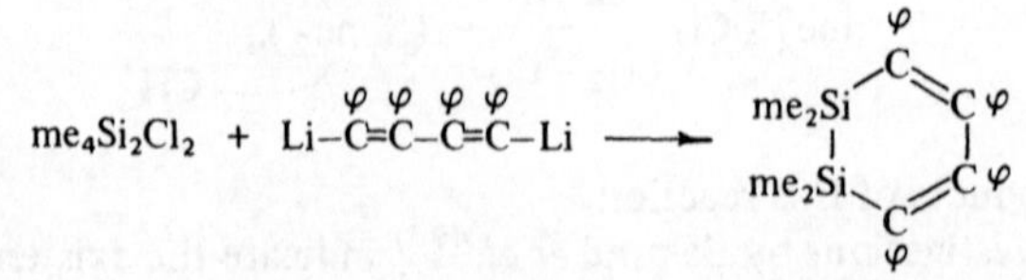

The UV data indicate nonplanar geometry in the diene chromophore.

Chernyshev *et al.* found an unsaturated C-containing ring:

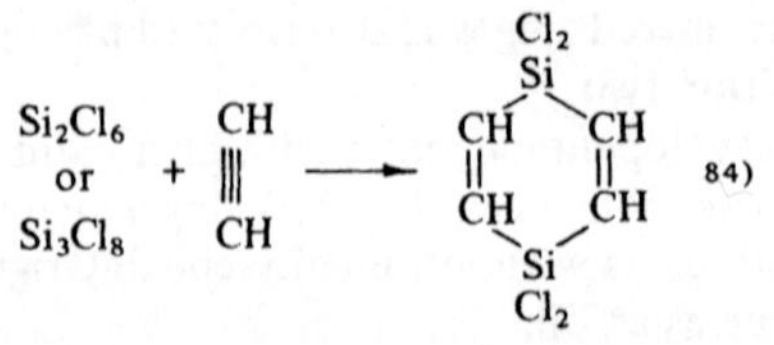

Other group-IV elements rarely occur in heterocyclic silanes. Germanium, the element most similar to silicon, forms two known ring systems, one being a five- and the other six-membered ring. The syntheses however, are different. Carberry and Dombek[73)] obtained two six-membered ring systems from the reaction of dimethylgermanium dichloride and dimethylsilicon dichloride with Li in kind of a Wurtz synthesis:

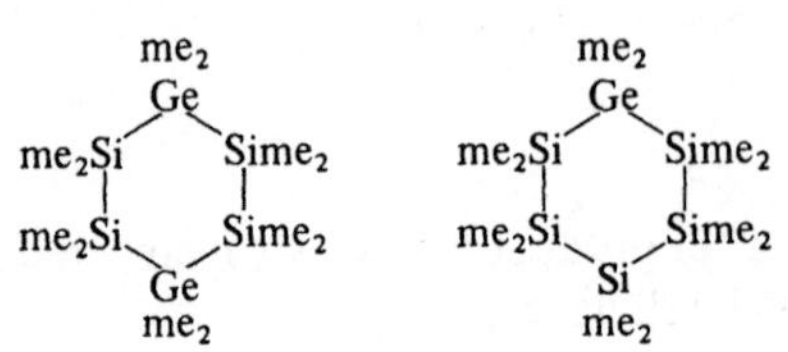

The five-membered ring is formed by a ring closure reaction from a 1,4-dilithium octa phenyltetrasilane with dimethylgermanium dichloride[256]:

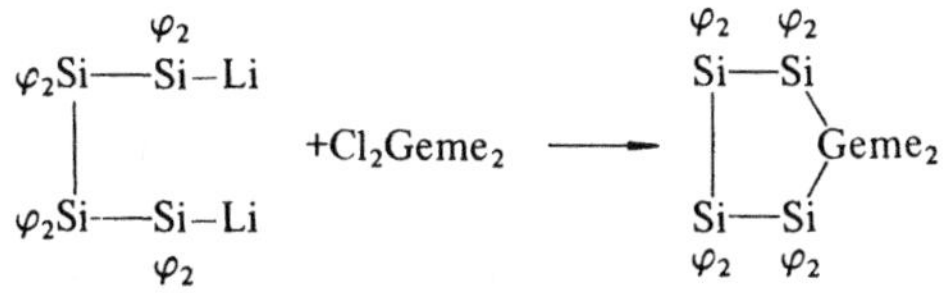

Even a tin derivative was observed after a reaction of the same type, *i.e.* the reaction with dimethyltin-dichloride[256].

Heterocycles with elements of group V are known too, chiefly with nitrogen. Only a few compounds with phosphorus have been described. Nonaphenylphosphocyclopentasilane is formed by the reaction described above[256]:

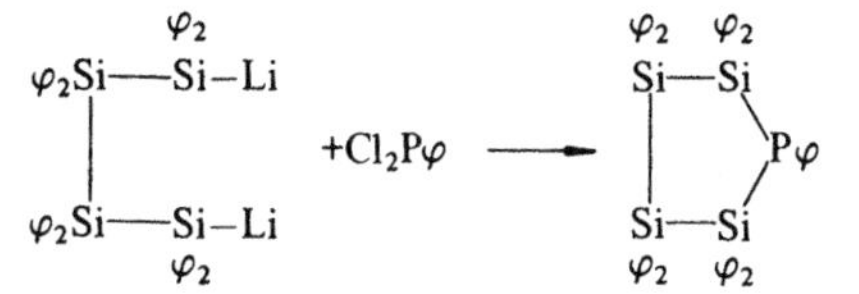

A four-membered ring with two phosphorus and two silicon atoms was found[294] at the reaction of

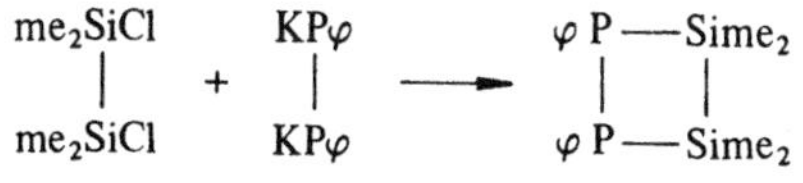

N-containing heterocyclosilanes were formed in various kinds. The first step, cyclization of the di-iodide I $(Si\varphi_2)_4$I with amines, yields in a first step a chain with NR_2 groups. At higher temperature, a second step and a cyclization occurs:[256, 281]

$$I(Si\varphi_2)I \;+\; 4H_2Nme \longrightarrow me\,HN(Si\varphi_2)_4NHme \;+\; 2NH_2me.\,HI$$

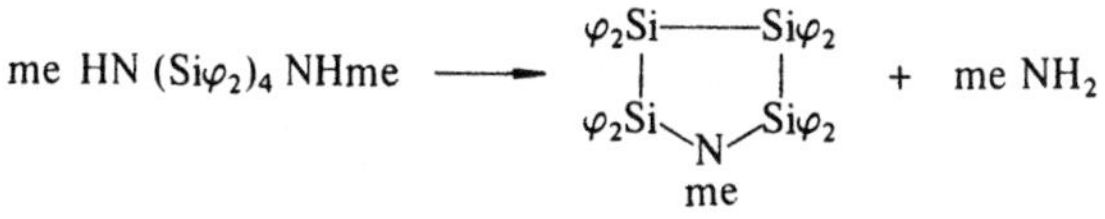

Two nitrogen heterocycles involving Si-Si bonds are known from the work of Wannagat and Brandstätter. A five-membered ring was formed by the interaction of 1,2-dichlorotetramethyldisilane and dilithium bis-(methylamino)-dimethylsilane[655, 658]:

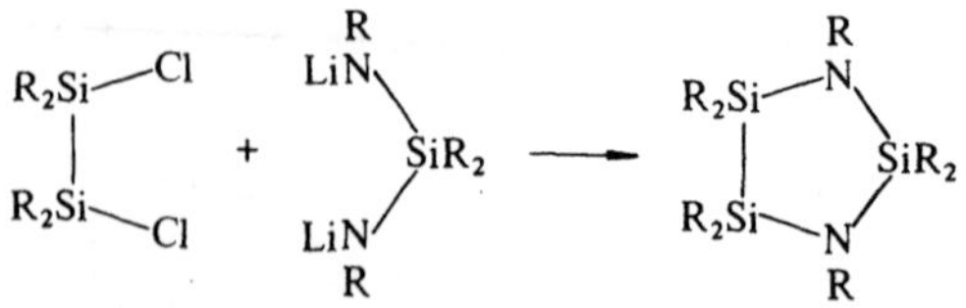

and six-membered rings with R=H and CH_3 are the products of the solvolysis of 1,2-dichlorotetramethyldisilane with ammonia or methylamine[656]:

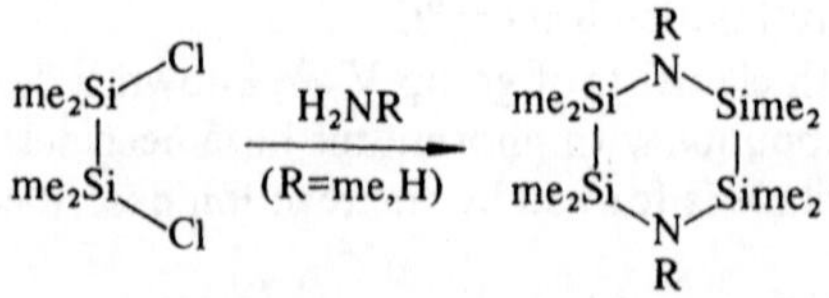

It is surprising that the mole refraction of the latter two compounds is not in agreement with the rules of Eisenlohr and Lorenz-Lorentz. No explanation has so far been given for the difference between experimental data and the theory. The Si–Si bond has not been found to have an abnormal increment.
A similar six-membered ring was found by the reaction of 1,2-dichlorotetramethyldisilane with hydrazine[89]:

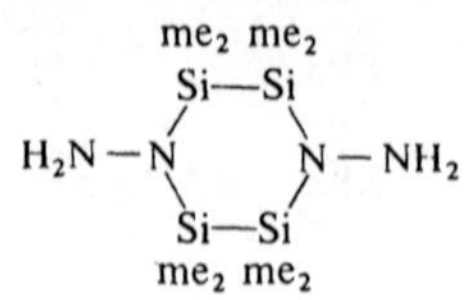

The corresponding trisilane $Si_3(CH_3)_7Cl$ yields only a nonvolatile residue. The analytical and molecular weight data obtained indicate the formula $[Si_3(CH_3)_6N_2H_2]_5$. A heterocyclic structure is probable.

Wannagat *et al.*[660] recently produced the following new N-containing heterocyclosilanes (mostly permethylated) by the general reaction of N-lithium compounds with chlorosilanes:

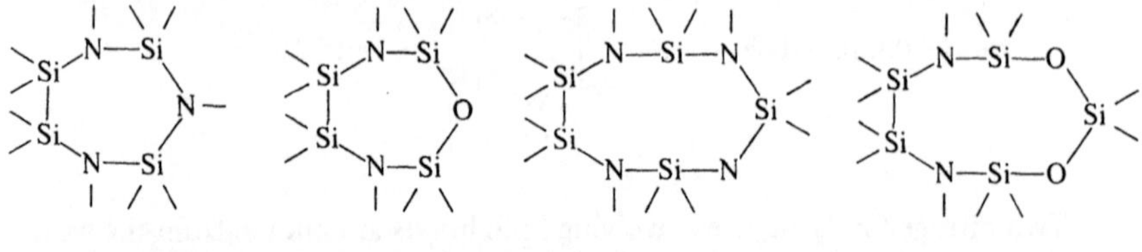

The oldest known heterocycles are those containing oxygen as the hetero element. Kipping's work[386), 382] with the first cyclosilanes (formed by a Wurtz synthesis from diphenyldichlorosilane) led to the discovery of two oxygen-containing heterocycles. Kipping called these compounds the "monoxide" and the "dioxide". The monoxide was obtained by hydrolysis of 1,4-diiodooctaphenyltetrasilane. Recent investigations have confirmed that this postulated structure is an octaphenyl-oxa-cyclopentasilane. It can also be formed hydrolyzing 1,4-diacetoxyoctaphenyl-tetrasilane(produced by reaction of octaphenylcyclotetrasilane with mercuric acetate[178]):

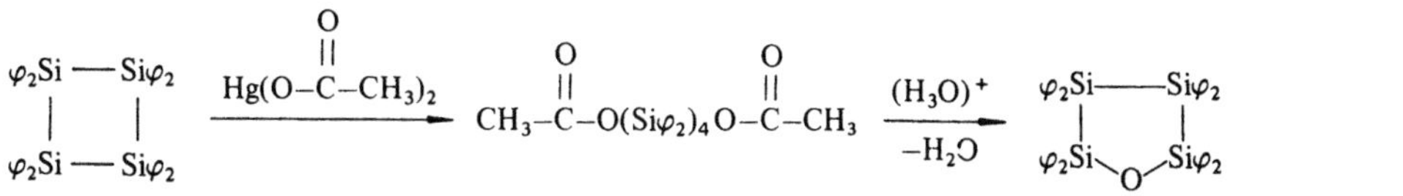

The IR spectrum of this octaphenyloxacyclopentasilane showed a very intense band at 955 cm^{-1}, an abnormally low frequency for the siloxane group, but more recent measurements[281] show the band at 988 cm^{-1}.

The dioxide is known in two isomeric forms. Kipping obtained a "rhomboidal oxide" by oxidation of the monoxide with nitric acid or moist iodine. The oxidation of octaphenylcyclotetrasilane with nitrobenzene yielded the other isomeric dioxide.

The tentative structures proposed for these two compounds are[318]:

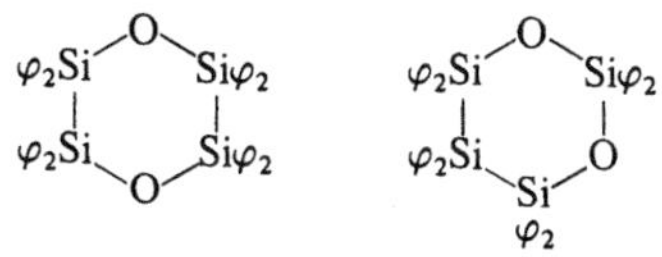

I II

Kipping's "rhomboidal oxide" has the structure I as demonstrated by Gilman by the condensation of two molecules of sym-tetraphenyldisilanediol. The other isomer seems to have the structure II.

Phenyl derivatives of oxygen-containing heterocycles are not the only known derivatives. Kipping described two five-membered rings with two other substituents, p-tolyl[592] and benzyl groups[591].

More recent work has been done by several Japanese chemists. In 1956, Kumada *et al.*[371] found a methylated six-membered ring with two oxygen atoms as the hetero atoms in a silicon ring after hydrolysis of 1,2-dichloro-tetramethyldisilane. The fluorine derivative gives the same reaction. The crystal structure was also investigated[613].

Linear, cyclic, and cage-like methylsilicon telomers having a framework of SiSiO were found by Kumada *et al.*[360]. Telomerization of 1,1,2,2-tetraethoxydimethyldisilane by acid cleavage with HCl in ether gives a cage-like telomer (II). Cohydrolysis of the same ethoxydisilane with $me_2(Oet)SiSi(Oet)me_2$ was effected by treating with HCl to give the mono-(I), the di-(III) and the tricyclic (IV) telomer:

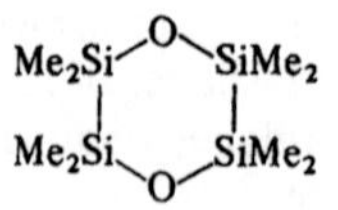

I

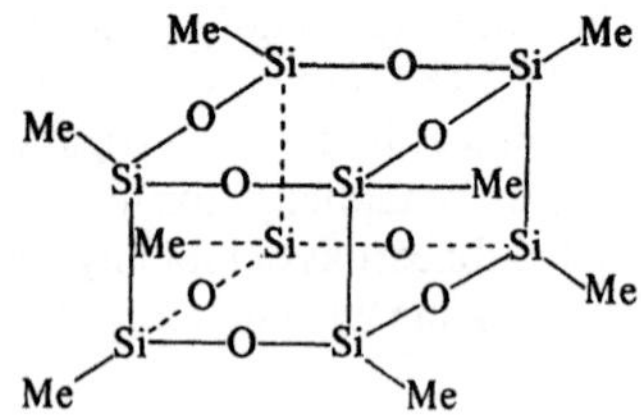

II

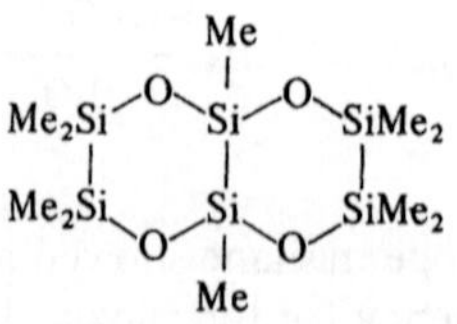

III

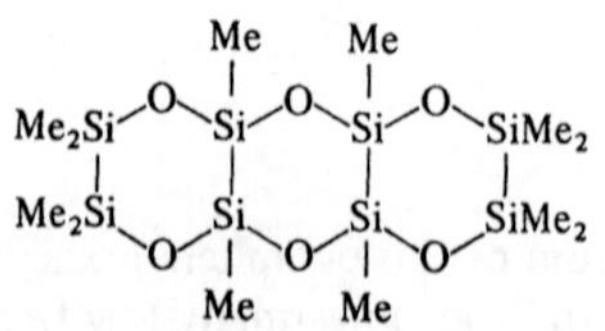

IV

Under comparable conditions $me_2(Oet)SiSi(Oet)me_2$ and $me_2Si(Oet_2)$ gave the expected heterocycles in a similar reaction:

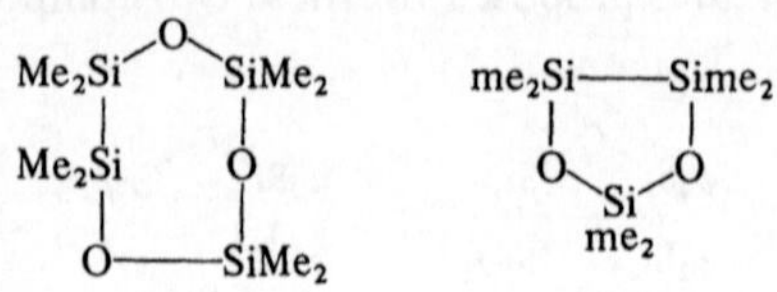

Crystal and molecular structures of II, I and IV were determined by Higuchi and coworkers[286, 287, 613], and were confirmed to be the structures described above.

Some heterocycles with sulphur exist. Dilithiumoctaphenyltetrasilane reacts with SCl_2 giving a small yield of a five-membered ring[256]:

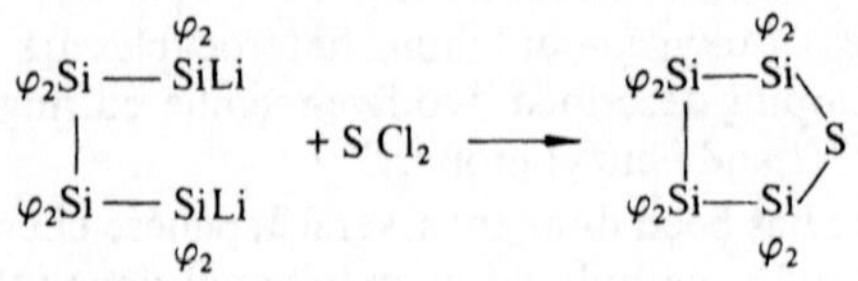

1,2-Dichlorotetramethyldisilane reacts with H_2S and forms a six-membered ring with two sulphur atoms[657]:

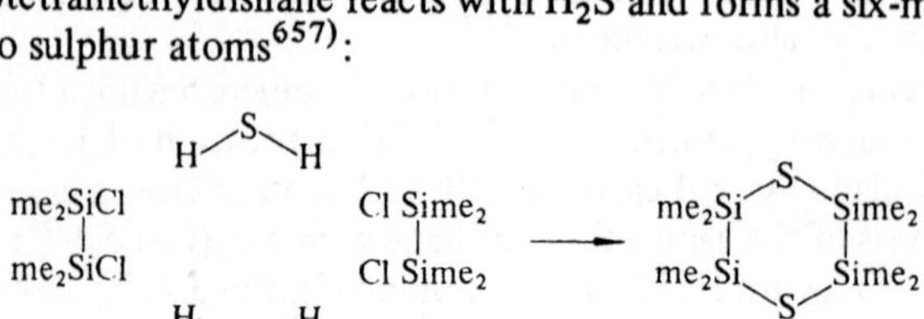

The first heterocyclic silane with boron as a hetero atom was recently prepared[280, 281]:

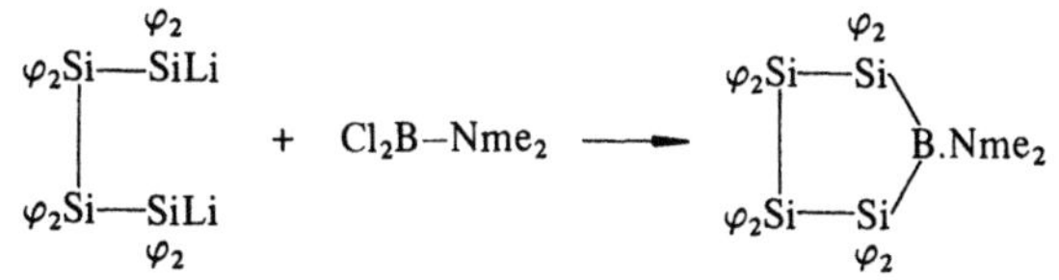

The compound is stable only with nitrogen as a substituent on the boron atom, which supplies electrons for stabilization of the B-Si bonds. Attempts to form a Si-B bond without a Nme_2 group were unsuccesful.

10. Polymeric Compounds with Si-Si Bond Systems

10.1. Polymers with Small Si–Si Units

There are some silicon compounds with small Si-Si units linked together by Si-O-Si, Si-NH-Si or similar bond systems. In many cases, such compounds are obtained by the solvolysis of oligosilanes with a subsequent condensation reaction. However, preparation from polymeric starting material is also possible, for instance, the formation of siloxene from calcium silicide.

The oldest publications in this field are mainly concerned with the hydrolysis of hexachlorodisilane and octachlorotrisilane[166, 167, 417, 418, 575]. The chemical and physical properties of the compounds are well described, but it now seems necessary to revise the postulated structural formulas and the nomenclature (*e.g.* siliconoxalic acid and siliconmesocalic acid). Exact structural formulae have yet to be ascertained, but the structure certainly includes Si_2 and Si_3 units condensed with Si-O-Si bonds. Depending upon the reaction conditions and water content, more or fewer OH groups will be found in the compound. However, in no case are the double bonds $>Si=Si<$ or $>Si=O$, which had been postulated in earlier publications, found to be present.

The described properties are in good agreement with the current concept of the structure. The compounds are white amorphous powders, insoluble in all solvents except by decomposition. With aqueous ammonia or caustic alkalis, hydrogen is evolved. With abrupt mechanical or thermal shock, explosive decomposition may occur, in which the enthalpy of oxidation of the Si-Si bonds to Si-O bonds is released.

10.1.1. Preparation by Solvolysis of Silicon Halides

A similar interpretation may be based upon the results of Schwarz and coworkers[580, 581]. These authors hydrolyzed higher silicon chlorides and found solid white polymeric compounds in which the original Si-Si framework is retained. An irregular network containing Si-O-Si bond systems is also formed, so that the earlier postulated structural formulas are certainly incorrect.

Recent results in this field were reported by Schumb *et al.*[575, 576]. Careful partial hydrolysis of hexachlorodisilane yields a series of oxyhalides: Si_4OCl_{10},

$Si_6O_2Cl_{14}$ and $Si_8O_3Cl_{18}$. The isolation of these colorless liquids is consistent with the postulated basic structures of the polymeric compounds.

Some results are known for the ammonolysis of oligochlorosilanes. After a first paper by Schwarz and Sexauer[589)] more recent results were given by Billy[38–41)]. Si_2Cl_6 reacts with ammonia to $Si_2(NH_2)_6$ (not isolated) which immediately evolves ammonia and forms a polymer $[Si_2(NH)_3]_n$. On heating this compound, a nitride, Si_2N_2, will be formed. All these compounds are sensitive to decomposition by hydrolysis, by which they are converted to silicic acid, hydrogen and ammonia.

10.1.2. The Chemistry of Siloxene

We have seen in the first chapter that silicon is not able to form double bonds with itself. In the second period of the periodic table, an overlap between two *p* orbitals is impossible because the distance between two atoms is too great. A more sophisticated picture of the Si-Si system shows that bond strengthening can occur via *d* orbitals, and we have discussed these effects (see page 53).

In the nineteenth century, when the picture was still rather crude, chemists attempted to prepare compounds with a "true" Si-Si double bond. An over-simplified analogy was relied on at that time: carbon-carbon multibonds are formed during the decomposition of carbides like calcium carbide, therefore compounds with Si-Si multiple bonds ought to be formed in the decomposition of calcium silicide.

Several chemists tried this approach. The most important paper of this era was published by Friedrich Wöhler[687)] in 1863. It is exciting to read this paper even today.

CaC_2 yields the gas C_2H_2 on reaction with acids, while decomposition of $CaSi_2$ by acids produces a yellow solid composed of silicon, oxygen and hydrogen. Wöhler named this first colored silicon product "Silicon". This has inadvertently caused some confusion between German and English nomenclature (silicon vs. Silicium, not to mention silicons and Silicone). In German, the silicones and Wöhler's "Silicon" have the same name. There is no word in English for Wöhler's "Silicon" and because it has no similarity with polysiloxanes, I will from now on call it the "Wöhler compound".

In his first paper, Wöhler does not discuss any structural aspects of this compound. One of the striking properties of the Wöhler compound is its yellow color and strong yellow fluorescence on irradiation with UV light. These very properties prompted Kautsky[335)] to start new investigations in 1921. He found that the Wöhler compound changes its properties with the conditions of preparation and is actually a mixture of several different compounds.

Milder and better controlled reaction conditions resulted in a white compound which burns spontaneously upon exposure to air, and is also sensitive to water, caustics, and to many other reactants.

Kautsky[342)] called it "siloxene" and found the structure to be a polymeric layered compound with six membered silicon-rings linked by oxygen. The fourth valency of the silicon is occupied by hydrogen. The ring is not planar but rather irregular, so that the fourth valency of each silicon sticks up out of the layer: three bonds of every ring face up, and three bonds down.

Siloxene has two remarkable properties: all the hydrogen atoms are exchangeable with other atoms or groups and the whole compound has the appearance of a two-dimensional layer; and the characteristic colors and the fluorescence of such derivatives are significant.

A few more details about these two properties will enable us to understand why Kautsky proposed this structure. The starting material is $CaSi_2$, a layer lattice with silicon and calcium layers one upon another. This structure was elucidated by X-ray analysis carried out by Böhm and Hassel[43]. Weiss[316] *et al.* secondly repeated this work using more refined techniques. These papers differ only in some minor aspects of layer-stacking. We consider that only the silicon layer is important. It contains six-membered siliconrings condensed togehter. The rings are irregular as described above.

$CaSi_2$ yields siloxene via acid hydrolysis (HCl) in darkness in alcoholic solution under nitrogen. Siloxene is a white powder which is insoluble in all solvents, gives no X-ray lines, burns spontaneously in air and evolves hydrogen with caustics. The structure was shown above.

The Si-H bonds are extremely reactive and the hydrogens can be replaced by halogens without destroying the siloxene structure, forming the siloxene halide derivatives.

To emphasize the layer structure and show why hydrogen could be exchanged without disrupting the layer, Kautsky[336] drew the part of formula which represents the layer surrounded by a rectangle. All exchangeable atoms are then outside this box:

$$\boxed{Si_6O_3}\ H_6$$

hence no reaction can occur with the atoms within the box without decomposition of the layer.

Bromination has been studied in the greatest detail[341, 342, 343, 349]. Gaseous hydrogen bromide reacts with siloxene to form monobromosiloxene; further bromination with HBr is not feasible:

$$\boxed{Si_6O_3}\ H_6 + HBr \longrightarrow \boxed{Si_6O_3}\ {}^{H_5}_{Br} + H_2$$

In contrast, elemental bromine reacts with siloxene and, depending on the reaction conditions (temperature, time) and the bromine concentration, gives bromosiloxenes of varying degrees of substitution.

Dibromosiloxene is the first product which can be isolated. It is formed in the reaction between a carbon disulphide solution of bromine (about 10%) and a carbon disulphide suspension of siloxene; the reagents are mixed while cooling with ice, as in a titration. When the bromine color remains, a dibromosilo-

$$\boxed{Si_6O_3}\ H_6 + 2\,Br_2 \longrightarrow \boxed{Si_6O_3}\ \genfrac{}{}{0pt}{}{Br_2}{H_4} + 2\,HBr$$

$$\boxed{Si_6O_3}\ \genfrac{}{}{0pt}{}{Br_2}{H_4} + Br_2 \longrightarrow \boxed{Si_6O_3}\ \genfrac{}{}{0pt}{}{Br_3}{H_3} + HBr$$

$$\vdots$$

$$\boxed{Si_6O_3}\ \genfrac{}{}{0pt}{}{Br_5}{H} + Br_2 \longrightarrow \boxed{Si_6O_3}\ Br_6 + HBr$$

xene with the stoichiometric ratio Si : Br = 6 : 2 has been formed. In contrast, when concentrated bromine solutions are allowed to act on siloxene suspensions at room temperature for prolonged periods, higher bromine derivatives are formed in which the degree of substitution is generally non-uniform. The maximum attainable Si : Br ratio is 1 : 1, corresponding to the fully brominated hexabromosiloxene, $Si_6O_3Br_6$.

The dibrome derivative stage is the most probable for steric reasons (Kautsky and Siebel[349]). From X-ray measurements on $CaSi_2$, it may be assumed that the Si_6 ring in siloxene is similarly angular (see Fig. 2), with three valencies directed towards one side of the laminar plane and three valencies towards the other. According to these authors, a bromine atom can readily be accommodated on each side of the Si ring. Further bromination, which requires the accommodation of at least two Br atoms on the same side of the ring, is sterically more difficult and consequently requires more time and a higher bromine concentration.

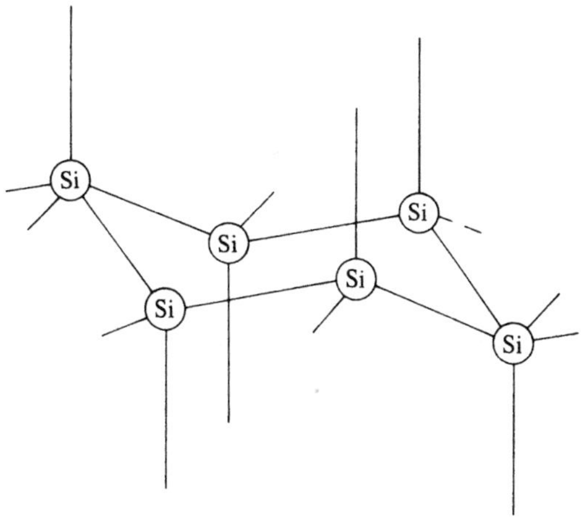

Fig. 2. Si_6 ring in siloxene and in $CaSi_2$.

The above arrangement in the dibromosiloxene molecule and the angular nature of the ring have been confirmed by recent investigations by Hengge and Grupe[258].

Halogen siloxenes usually react with alcohols to form the corresponding alkoxysiloxene derivatives:

$$\boxed{Si_6O_3}{}^{Br_x}_{H_{6-x}} + x\,ROH \longrightarrow \boxed{Si_6O_3}{}^{(OR)_x}_{H_{6-x}} + x\,HBr$$

This alcoholysis can also be carried out with diols, bridge formation across the ring being feasible under certain conditions.

In the case of diols such as n-propane1,3-diol, dibromosiloxenes react with only one OH group, in accordance with the angular ring structure in which one bromine atom is on one side and the second on the other side of the ring. Bridge formation is therefore impossible.

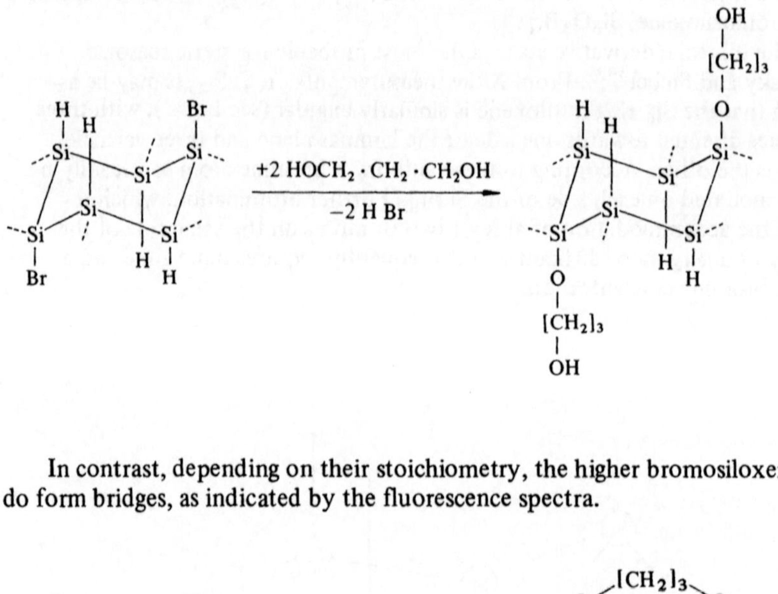

In contrast, depending on their stoichiometry, the higher bromosiloxenes do form bridges, as indicated by the fluorescence spectra.

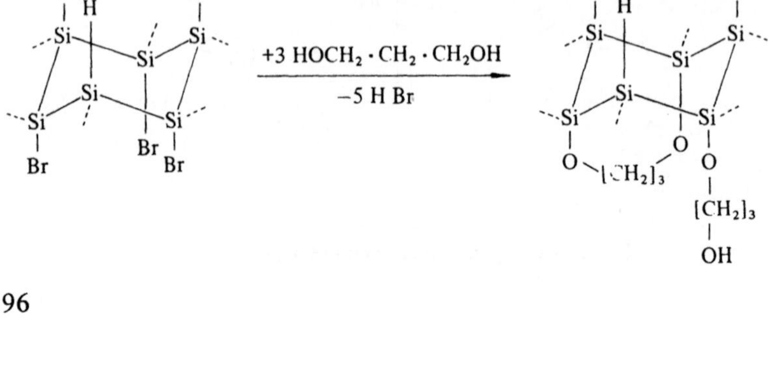

In this case, at least two bromine atoms lie on the same side of the ring, so that bridge formation can take place with suitable diols.

Few studies have been published on the iodosiloxenes[343, 348]. Owing to the greater size of the iodine atom, only the diiodosiloxene can be formed. The higher iodine derivatives are excluded for steric reasons. With elemental iodine, siloxene forms only a monoiodosiloxene $\boxed{Si_6O_3}{}^{I}_{H_5}$. Diiodosiloxene can be prepared indirectly by the reaction between bis-(diethylamino)siloxene (obtained from dibromosiloxene via solvolysis with ethylamine) and treatment with hydrogen iodide:

$$\boxed{Si_6O_3}{}^{Br_2}_{H_4} + et_2\,NH \longrightarrow \boxed{Si_6O_3}{}^{(Net_2)_2}_{H_4} + 2\,HBr$$

$$\boxed{Si_6O_3}{}^{(Net_2)_2}_{H_4} + 4\,HI \longrightarrow \boxed{Si_6O_3}{}^{I_2}_{H_4} + 2\,HNet_2 \cdot HI$$

Iodination of siloxene with organic halogen derivatives is also possible, particularly under the action of light.

Chlorine derivatives of siloxene are also known. The action of elemental chlorine on siloxene is much more vigorous than that of bromine or iodine. In addition to the Si–H bond, the Si–Si bond is also attacked to effect decomposition of the siloxene network. One of the decomposition products is $Cl_3SiOSiCl_3$[334]. No fluorine derivatives of siloxene have as yet been described, but many other groups exchange with halogen or hydrogen, *e.g.* OR, NHR, and SR groups.

The OH derivatives are very unstable and the black hexahydroxysiloxene is explosive in the dry state[351]. In all these cases there are from one to six substituents in each ring. The only limitation on substitution is that there must be sufficient space for the group on the layer.

Other derivatives of siloxene were also prepared[348]:

$$\boxed{Si_6O_3}{}^{H_4}_{(OH)_2} \xrightarrow[(me\,CO)O]{me \cdot COOH\ or} \boxed{Si_6O_3}{}^{H_4}_{(OCOCH_3)_2}$$

$$\boxed{Si_6O_3}{}^{H_4}_{(Ome)_2} \xrightarrow{me \cdot COOH} \boxed{Si_6O_3}{}^{H_4}_{(OCOCH_3)_2}$$

$$\boxed{Si_6O_3}{}^{H_4}_{(NHet)_2} \xrightarrow{me \cdot COOH} \boxed{Si_6O_3}{}^{H_4}_{(OCOCH_3)_2}$$

Instead of starting with the unstable hydroxysiloxene, one can start with alkoxy or alkylaminesiloxenes and acetic acid, along with other carboxylic acids or acid anhydrides.

One may also react SO_2, CS_2 and CO_2 with the amino derivatives to form amido acids[348]:

$$\boxed{Si_6O_3}\begin{matrix}H_4\\(NHet)_2\end{matrix} + CO_2 \longrightarrow \boxed{Si_6O_3}\begin{matrix}H_4\\(NHet\text{-}COO^-)\ (NH_2et)^+\end{matrix}$$

$$\boxed{Si_6O_3}\begin{matrix}H_4\\(NHet)_2\end{matrix} + SO_2 \longrightarrow \boxed{Si_6O_3}\begin{matrix}H_4\\(NHet\text{-}SOOH)_2\end{matrix}$$

Li Alkyls react with siloxene to form a dark product of the composition[338]:

$$\boxed{Si_6O_3}H_6 + 2\,(n+m)\,LiR \longrightarrow \boxed{Si_6O_3}\begin{matrix}H_{6-(n+m)}\\R_n\\Li_m\end{matrix} + nLiH + mRH$$

$$n \approx m$$

An interesting aspect from the view point of organic chemistry is the reduction of ketones and aldehydes by siloxene[345]:

$$\boxed{Si_6O_3}H_6 + R^1R^2CO \rightarrow \boxed{Si_6O_3}\begin{matrix}H_{6-n}\\(O\text{-}\underset{H}{C}R^1R^2)_n\end{matrix}$$

$$\xrightarrow{H_2O} \boxed{Si_6O_3}\begin{matrix}H_{6-n}\\(OH)_4\end{matrix} + R^1R^2CHOH$$

The carbonyl group reacts with the Si-H bond and subsequent hydrolysis forms the alcohol.

All these exchange reactions occur rapidly and quantitatively. There is no hindrance due to diffusion, in the solid state. The entire surface is capable of reacting. Chemisorption with bromine permits measurement of the surface area ($5.9 \times 10^5\,m^2/mol$).

Kautsky and Pfleger[347] measured the surface by the BET method and found only about 30% of the theoretical surface when using CH_4, and still less with nitrogen; it appears that these very small bond interactions are not able to separate the layers.

The best explanation of the above facts may be as follows. The layers of the $CaSi_2$ are separated by the action of the halogen hydride and the Ca layer is eluted in the form of Ca chloride. Hydrogen atoms bond to the Si layer whose negative charge becomes smaller, weakening the layer structure. Oxygen from water now reacts with the layer and the hydrogen of the water is reduced to gaseous H_2. The siloxene layers now lie loosely upon one another like the pages of a book and are bound together only be weak van der Waals forces. Compounds which react with the hydrogen exothermically are able to separate the layers, whereas the absorption of gases N_2 or CH_4 can only partially effect this.

Kautsky's postulation of a regular insertion of oxygen in the silicon layer forming six-membered Si heterocycles is based on the following considerations:

1. the lowest substitution ratio of all substituents, proved with bromine and iodine, is Si : substituent equals 6 : 1.
2. six-membered Si rings are preformed in the silicide;
3. electron microscopic pictures of siloxene show hexagonal particles[352]. All corners have angles of 120°.

The six-membered ring is highly probable, but it seems desirable to have still more evidence for it.

The second outstanding property of the derivatives of siloxene is that they are colored and fluorescent. Siloxene itself is greenish white and exhibits slight blue-green fluorescence. The absorption spectrum has a maximum at 380 mμ. Investigations of these features gave the following picture[252, 271]:

1. The absorption spectra and the fluorescence spectra are mirror-images of each other. The duration of the fluorescence is of the order of 10^{-8} seconds. The fluorescence spectra are insensitive to small impurities and small changes in the preparation methods. The conclusion is that the color and the fluorescence are properties of the molecule and not of the lattice[272], so the color must originate from the peculiar type of bonds in the system.
2. The fluorescence maxima shift toward longer wavelengths when the number of the substituents is increased. We have carried out investigations with substituents like halides, OR, NHR etc. having free electron pairs.
3. Both the numer and type of substituents influence the magnitude of the shift.
   An investigation of several substituents gave the following series with an increase of the shift to longer wavelengths:

$$\text{Br} < \text{Oet} < \text{Ome} < \text{I} < \text{SR} < \text{NHet} < \text{OH}.$$

The effect of these substituents on the color of polymeric compounds such as $(SiX)_n$ and $(SiX_2)_n$ is nearly the same. An increase in the force constants of disilanes was also observed within a similar substituent series.

The origin of the color appears to be the same in each case. Therefore, one can say that whenever some Si atoms are directly bonded together and also have a high electron density, the compound will be colored. This is a result of electron-donating groups bonded to siloxene. Where the electron density of a silicon atom in the Si_6 ring is increased by additional dπpπ bonds of the substituents, this increased electron density leads to additional links between the silicon atoms of the ring in siloxene. The oxygen atoms in the layer reinforce this effect.

These additional bonds, which presumably are no longer localized, account for the chromophoric properties of the Si ring. This view is supported by the fact that a hypsochromic shift occurs when the free electron pairs of a substituent are blocked, for example, in an OR group by addition of BF_3.

Why do siloxene and its derivatives exhibit fluorescence in contrast to all other colored compounds encountered in silicon chemistry? So far, no investigations have been reported, but we can make some plausible conjectures. Fluorescence normally occurs if the absorbed radiant energy is not destroyed in a non-radiative process such as occurs in isolated centers of a molecule, in a lattice, or in a very rigid arrangement.

In siloxene, isolated Si rings probably form a center "insulated" by oxygen and it is this that causes the fluorescence.

It is perhaps relevant here that the fluorescent compound formed in the hydrolysis of iodine cyclosilanes also has rings isolated by oxygen. It is significant, moreover, that two-dimensional polysilanes become fluorescent if a slight partial oxidation takes place.

Let us return now to the Wöhler compound. What light does the chemistry of siloxene shed on the structure of this compound? The Wöhler compound is formed from $CaSi_2$ with HCl in high concentration in light and air. Under these conditions, siloxene forms halogen derivatives which react with water to hydroxysiloxenes.

Hydroxysiloxenes are intensely colored, highly fluorescent and very sensitive to light. The colors shift to lower wavelengths under the influence of light, but only as far as a certain limiting wavelength. In the fluorescence spectra of the methoxysiloxenes, the same effect occurs, but more slowly. Once the limiting wavelength is reached the extinction of the color and the intensity of fluorescence declines. This limiting wavelength is the same as the wavelength of maximum fluorescence for the Wöhler compound. Therefore, the Wöhler compound seems to be identical with a bleached hydroxysiloxene[269].

The chemistry of the bleaching reaction is not known in detail, but some evidence indicates that the main reaction is the rearrangement:

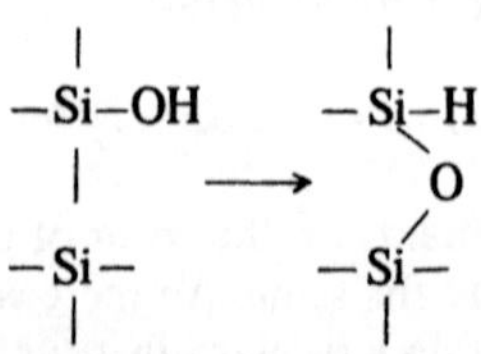

This new Si-O-Si bond is probably formed between the layers, making an irregular three-dimensional network.

Finally, the Wöhler compound fluoresces with a yellow color. When oxidized with a strong oxidizing reagent like $KMnO_4$, its oxidation energy is large enough to elevate the native part of the molecules into an excited state, when fluorescence takes place. This is a chemiluminescence phenomenon. The spectra are identical with the fluorescence spectra[346, 353].

Now we know that not every oxidation reaction in the presence of fluorescent material produces fluorescence. The systems are not coupled, and there is no energy transfer between the first reaction and the fluorescent material. So it seems possible that energy transfer may occur in the silicon layer itself, resulting in oxidation in one part of the layer. The resulting energy transfered through the layer produces an excited state in a native (*i.e.* yet unoxidized) six-membered ring system, and hence fluorescence occurs.

There is another surprise in this system. If a dyestuff is applied to the Wöhler compound, there are two possibilities:

1. the dyestuff is adsorbed on the surface of the Wöhler compound, whereupon oxidation of the system causes the dyestuff to fluoresce (when its excited state is lower than that of the Wöhler compound);
2. the dyestuff is not adsorbed, and no fluorescence takes place.

This phenomenon may be considered as sensitized fluorescence. Direct contact is necessary for this phenomenon to take place. Siloxenes and similar compounds have many very remarkable properties. Not all observed effects can be explained and some explanations are presently tentative or speculative.

10.2. Compounds with Polymeric Si–Si Systems

10.2.1. Introduction

Polymeric compounds with limited Si-Si units in which the polymeric character is due to other bond systems, have been described. Another series of polymeric compounds exists, in which the Si-Si bonds themselves are responsible for the polymeric state. The extreme case, when all four valencies of a silicon atom are bonded with Si atoms, produces metallic silicon with a diamond structure. Compounds of the type $(SiX_2)_n$ or $(SiX)_n$ result when some valencies of each silicon atom are occupied by other atoms or groups. Only polymeric compounds $(SiX_y)_n$ with y from 2 to 1 exist; SiX_3 yields disilanes, and compounds with y between 3 and 2 yield limited chains. The possible stoichiometric compositions SiX_2 and SiX are sometimes found, but a non-stoichiometric composition is more common, which is understandable in view of the irregular framework of the Si-Si structure. The formation of regular or irregular Si-Si structures of stoichiometric or non-stoichiometric composition will depend on the procedures used in preparation.

These polymeric compounds can be classified in two groups according to the method of preparation:

1. Compounds formed from low-molecular-weight silicon compounds by thermal polymerization reactions. Normally, an irregular polymerization takes place, resulting in products with various compositions. Compounds of defined composition sometimes result, but need not necessarily have a regular structure, *e.g.* when the monomeric starting material forms radicals of defined composition such as in the formation of "silylenes". The usual assumption that chains and rings are formed exclusively by the polymerization of SiX_2 units is not always correct. A rearrangement sometime occurs:

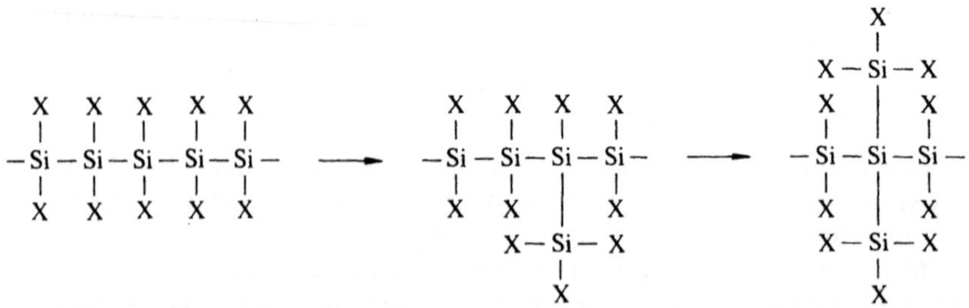

Irregular structures are then formed, even with a stoichiometric composition. The picture becomes more complicated due to formation of other monomeric radicals (*e.g.* SiX) or secondary reactions, for instance a reaction of excess alkali metal with ready-formed $(SiX_2)_n$. Small differences in the formation conditions affect the products, which explains why there are differences between the results of different authors.

2. A second group of compounds of the type $(SiX)_n$ will be formed from polymeric starting material in which the Si-Si frame work is pre-formed. A series of compounds can be formed from some silicides by exchange of the substituents without changes in the Si-Si framework. Thus polymeric compounds of well-defined structures and stoichiometric composition are obtained. Stoichiometric composition may be affected by steric hindrance of substituents in the exchange. Such complications will be described later.

The two groups of substances thus have completely different structures and totally different paths leading to their formation. A similarity in the chemical properties is due to similarity of the chemical bonds. Some physical properties are also similar, such as color which is due to Si-Si bond. Each group of compounds will be described separately.

10.2.2. Compounds with Polymeric Si–Si Bond Systems Obtained from Low-Molecular-Weight Starting Material

Polymeric Si-Si compounds can be formed by thermal or chemical polymerization of monomeric Si compounds, as already described. Some investigations indicated the appearance of intermediate radicals such as silylenes (see the review of Bürger and Eujen[67a]) and cf. p. 58) or others, which were sometimes isolated or identified spectroscopically. Recombination of these reactive units results in more or less irregular polymeric compounds.

Formation of monomeric radicals can be brought about by

a) physical energy (pyrolysis, silent electrical discharge, light etc.). The resulting equilibrium:

$$SiX_4 \rightleftharpoons SiX_2 + 2\,X$$

can shift to the right by capturing the halogen radicals with hydrogen, or by recombination, etc.

b) disproportionation:

$$Si_2X_6 \rightleftharpoons SiX_4 + SiX_2$$

caused by heating, or the catalytic influence of tertiary amines.

c) the reaction of metals:

$$SiX_4 + 2\,Me^I \rightleftharpoons SiX_2 + MeX \quad \text{Me: Alkali, Mg etc.}$$

Two different types of reaction c) are possible:

Firstly, the reaction with alkali metals, magnesium or other similarly reactive metals, mostly in ethereal solution. The tendency toward formation of the metal halide is the probable driving force of the reaction. Secondly, the reaction with silicon metal at about 1000 °C. This type is called a transport reaction because silicon metal is transported in the gaseous phase:

$$SiX_4 + Si \rightleftharpoons 2\,SiX_2 \rightleftharpoons (SiX_2)\ \text{polym.}$$

In specific reaction conditions, it is possible to isolate SiX_2 in the polymeric state.

Although Schäfer and Nickl[543)] did not find any abnormal increase in the vapor density on heating pure $SiCl_4$, a spectroscopic investigaion[676)] has indicated the formation of $SiCl_2$ radicals, though in very small concentrations. It is unclear whether silicon metal is only the radical trapper of preformed SiX_2 radicals, or whether the first step is a reaction of silicon with $SiCl_4$.

The most important results with the various compounds may be summarized as follows:

Polymeric silicon hydrides: In 1880 Ogier[437)] described unsaturated yellow silicon hydrides obtained in a silent electric discharge in SiH_4. Stock and coworkers[605)] reported the formation of solid yellow hydrides during the photochemical decomposition of higher silanes. The hydrogen content of these amorphous compounds was between $SiH_{1.12}$ and $SiH_{1.22}$. Stock and Somieski found products $SiH_{1.1}$ up to $SiH_{1.7}$ when reacting SiH_2Cl_2, $SiHCl_3$ or Si_2Cl_6 with sodium/potassium alloy[600)] or sodium amalgam[543)]. Emeleus[33)] found similar solid hydrides in the thermal decomposition of di- and trisilanes.

In recent investigations Schott and Herrmann[561, 563)] attempted to prepare silicon hydride by reduction of tribromosilane with magnesium:

$$2\,SiBr_3H + 3\,Mg \longrightarrow \frac{2}{n}\,(SiH)_n + 3\,MgBr_2\,.$$

They found it impossible to form a silicon-hydride not containing small amounts of bromine by this method. A typical composition was $(SiH_{1.03}Br_{0.8})_n$. When prepared in this manner, silicon hydride is a lemon-yellow powder which exhibits a weak yellow fluorescence. It is hard and brittle. Optical anisotropy and the lack of X-ray diffraction lines indicate an amorphous structure. On heating in air, it ignites at 98–102 °C and is insoluble in all organic solvents. With water and ethanol, hydrogen is evolved:

$$(SiH)_n + 4nH_2O \longrightarrow n\,Si(OH)_4 + 4nH_2 .$$

Silicon hydride reacts with bromine or iodine in solution with discoloration and is a strong reducing agent. All these properties and detailed studies of the oxidation of $(SiH)_n$ led Schott and coworkers to conclude[562, 564] that the silicon hydride isolated by them has an irregular structure.

Polymeric silicon fluoride: Silicon fluorides in the polymeric state are products of the recombination of monomeric SiF_2 radicals. These SiF_2 radicals are described in detail in Chapter 7: Silylenes (see p. 58).

Various experimental studies were carried out with the object of obtaining polymeric $(SiF_2)_n$ without a preformation of monomeric units. When fluorination of $(SiBr_2)_n$ was attempted with fluorinating agents such as HF, SbF_3, AgF, AgF_2, AsF_3, and HSO_3F, in no case was $(SiF_2)_n$ obtained; only SiF_4 was observed[548]. However, fluorination of a layered $(SiF)_n$ is possible (see p. 109). This reaction thus seems to be a matter of the steric hindrance of the halogen exchange.

Formation of polymeric $(SiF_2)_n$ or $(SiF)_n$ is possible in the reaction of $SiBr_2F_2$ or $SiBr_3F$, respectively, with magnesium[360, 548]. Silicon monofluoride was also obtained on heating silicon difluoride to 200–350 °C. At higher temperature (400 °C) an explosive decomposition into silicon and SiF_4 occurs.

SiF_2 is also formed on thermal disproportionation of Si_2F_6[549]. Another disproportionation[553] occurs under the catalytic influence of triethylamine, forming a brown powder $(SiF_{0.91})_n$ which ignites spontaneously in air.

Polymeric silicon chlorides: One of the first detailed studies of the silicon chlorides was published by Besson and Fournier[34] in 1909. The authors found that a compound $(SiCl_2)_n$ was formed during silent electrical discharge in $SiHCl_3/H_2$ mixtures.

Later Schwarz and coworkers[580, 582, 584] reported the formation of higher chlorosilanes and of a yellow polymer chloride $(SiCl)_n$[579] in the pyrolysis of $SiCl_4$. The polymer is an amorphous yellow powder, stable up to 400 °C, but sensitive to water. On average, three valencies of every silicon atom are bonded to silicon while the fourth valence is occupied by a chlorine atom. The exact 1 : 1 ratio of Si : Cl is surprising, since a variety of compositions would be expected.

Schwarz and Pietsch[588] also described a white $(SiCl_2)_n$ of various compositions, obtained in an electric discharge in $SiCl_4/H_2$ gaseous mixtures. Wiberg and Hertwig[282, 283] recently repeated this work and found similar compositions, but in all other investigations on silicon chlorides in polymeric states a yellow color was found. The partial silicon-diamond lattice postulated by Wiberg lacks sufficient physical evidence.

Rochow and Diditschenko[474] found similar results, obtaining an oil, $SiCl_{2.61}$, by the reaction of $SiCl_4$ with silicon at 1000 °C.

The formation of monomeric $SiCl_2$ by the reaction of $SiCl_4$ with silicon was reported in Chapter 7 (p. 66).

Detailed investigations of the equilibrium and thermodynamic data were published by Schäfer *et al.*[539, 542, 543], Antipin[13], Wolf and a coworker[690], and Kempter *el at.*[378].

Under specific reaction conditions, $SiCl_2$ can be isolated in polymeric form, as was independently shown by Schenk[545] and Schmeisser[552]. The product isolated is soluble in organic solvents, and measurements of the molecular weight show a molecule composed of 16–17 $SiCl_2$ units. More detailed structural investigations have not been performed. Whether this compound has a polymeric structure or whether it is a chain or a ring of limited size is unclear.

Similarities of the $(SiCl_2)_n$ described here to compounds such as $Si_{10}Cl_{16}$, found by Schwarz *et al.*[581, 585, 590] are discernible.

Polymeric silicon bromides have a certain similarity to silicon chlorides, where an equilibrium between silicon and $SiBr_4$ also exists. Thermodynamic studies have been made[542, 688]. The isolation of polymeric $(SiBr_2)_n$ is possible in the form of a brown powder which dissolves well in benzene[551]. The molecular weight was estimated as approximately 3000. Above 200 °C a cleavage into $SiBr_4$ and $(SiBr)_n$ is observable[445]. $[Si(OH)_2]_n$ is unstable to hydrolysis and undergoes a strongly exothermic rearrangement, sometimes in an explosive reaction:

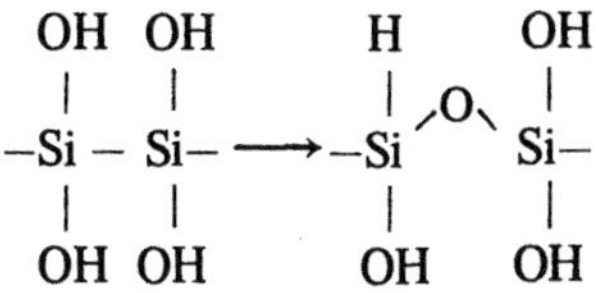

With $LiAlH_4$, a polymeric hydride $(SiH_2)_n$ is formed and with Grignardcompounds the corresponding alkyl compounds can be obtained:

$$(SiBr_2)_n + 2n\ RMgBr \longrightarrow (SiR_2)_n + n\ MgBr_2$$

$$(SiBr_2)_n \xrightarrow{LiAlH_4} (SiH_2)_n.$$

With magnesium, al slow reduction to $(SiBr_{1.46})_n$ occurs[551, 558]. Apparently a Si-Si framework is formed and the reduction becomes sterically hindered, so the reaction product has an irregular structure.

$(SiBr)_n$ formed in the reaction of magnesium with $SiBr_4$ seems to be identical with the bromide isolated after the pyrolysis of $(SiBr_2)_n$. Like all the other silicon halides, $(SiBr)_n$ can be formed by disproportionation of Si_2Br_6 in form of an insoluble yellow-brown solid. The particular reaction temperature confirms the postulated decrease of thermal stability in the series $Si_2Cl_6 > Si_2Br_6 > Si_2I_6$.

Polymeric silicon iodide: No such detailed investigations have been made of the iodides. An equilibrium between silicon and SiI_4 also exists, but the easy thermal decomposition of the iodides indicates a reversal of the temperature effects[542]. $(SiI_2)_n$ was found to be a brittle, amorphous, yellow-red solid[550]. Pyrolysis of SiI_4 yields a similar product, soluble in benzene. In an electric discharge, a yellow red solid is formed having the composition $(SiI_{2.2})_n$ and insoluble in all organic solvents. During pyrolysis at 220–230 °C, it reacts to form an insoluble product $(SiI_2)_n$ of dark red color in addition to SiI_4 and Si_2I_6. Pyrolysis of the product $(SiI_2)_n$ at 350 °C yields an orange colored $(SiI)_n$.

A silicon monoiodide is also formed by disproportionation of Si_2I_6[587]:

$$3\, Si_2I_6 \rightleftharpoons 2\,(SiI)_n + 4\, SiI_4.$$

On hydrolysis, an oxyhydride of the composition $(Si_2H_3O_4)_n$ is obtained, of which the structure is unknown.

The many different observed forms of silicon iodides in polymeric forms show the possibility and ease of thermal rearrangements and the existence of different irregular structures. The observed compositions seem to be more or less fortuitous.

Polymeric silicon alkyl compounds: Polymeric compounds are often observed in the Wurtz synthesis of dihalodialkyl(aryl)silanes with metals. The intermediate product is probably a dialkyl(aryl)silylene (see Chapter 7). A polymeric dimethylsilane was found by Burkhard[68]. A yellow polymeric $(SiCH_3)_n$ was isolated after pyrolysis of SiH_4 with ethylene[156].

In this connection the formation of irregular polymers $(Si\,Br_vH_yR_xO_z)_n$ with $v + x + y + z = 2$ is of interest. As a result of steric hindrance[566], such compounds will be formed in the reaction of $SiBr_4$ with t-butyl Grignard compound instead of the expected tetrakis (t-butyl)-silane.

10.2.3. Compounds with Polymeric Si–Si Bond Systems Formed by High-Molecular-Weight Starting Materials

The second principal possibility for forming of polymeric compounds with polymer Si-Si bond systems is via a substituent-exchange reaction starting from polymeric preformed Si-Si bond systems.

Investigations in this field have mainly used calcium silicides. In the Ca/Si system there are two silicides with Si-Si bonds in polymeric systems. In calcium monosilicide the silicon atoms are arranged in form of chains, while the calcium disilicide $CaSi_2$ has a layer lattice, consisting of silicon- and calcium layers. It is possible to form polymeric compounds $(SiX)_n$ and $(SiX_2)_n$ from both silicides by replacing calcium.

a) CaSi as starting material: In CaSi, silicon is arranged in zig-zag chains surrounded by Ca atoms[249]. Schwarz and Heinrich[583] reacted CaSi with alcoholic hydrochloric acid under inert gas and obtained a brown product with a Si : H ratio 1 : 1.67. A reaction in acetic acid yielded a product in which Si : H = 1 : 1.82. The expected polysilane $(SiH_2)_n$ could not be isolated, possibly due to the use of very impure calcium monosilicide. Royen und Rocktäschel[476] were able to use a purer CaSi and found a product $(SiH_{0.7})_n$ – $(SiH_{0,9})_n$ in a similar reaction. Recent investigations showed[279] that the reaction of pure CaSi with hydrochloric acid in alcoholic solution gives a product with a composition $[H_{0.9}Si(Oet)_{0.62}Cl_{0.09}]_n$. A deficit of substituents was also found here. This cannot signify the existence of free radicals, since the ESR spectra show only a very small paramagnetic signal. One may thus have to assume, as Royen suggested, a reaction between the chains, and the formation of a lattice.

The recent observations of a reaction of the alcohol with the silicon lattice are in good agreement with the results of Schott, who also found a reaction with alcohol and $CaSi_2$ under similar conditions (see p. 108).

CaSi does not react quantitatively with acetic acid in ethereal solution, probably because the Ca acetate formed has only a very small solubility in ether[279]. It yields products with Si-H and Si-OOC–CH_3 linkages. The IR absorption band of the Si-H bond is observable at 2100 cm^{-1}, which is a very low frequency. Such low-frequency absorption always occurs when the Si-H bond is vicinal to Si-Si bonds and is easy to chlorinate. In a fast, quantitative reaction with acetic acid in a solution of CCl_4, a product of composition $[H_{0.75}SiCl_{0.95}(OOC \cdot CH_3)_{0.4}]_n$ was obtained. This composition shows no deficit of substituents and no ESR signal is observable to support the existence of a chain structure.

Additional CaSi reactions were carried out with halogens, PCl_5, and other halogenating agents. In all cases a decomposition of the silicon chains to monomeric silane derivatives occurred before reaction of the calcium was complete, so no definable products could be obtained.

b) Calcium disilicide as starting material: Detailed studies have been made of the reactions of calcium disilicide, $CaSi_2$. X-ray diffraction studies by Böhm and Hassel and recent investigations by Weiss[316] show a layer lattice, with Si layers in the form of Si_6 ring condensed together in a layer. The rings are not planar but corrugated. The fourth valency of the silicon extends out of the layer. Three bonds of every ring face up and three bonds down.

Among the studies of Kautsky *et al.* on the formation of siloxene by reaction with aqueous hydrochloric acid (see Chapter 10.1.2.) were some on reactions of $CaSi_2$ that conserve the silicon layer.

Schott[567] attempted the reaction of $CaSi_2$ in the way like Schwarz had performed the reaction with CaSi with halogen halides in dry ethanol, but was unable to obtain a pure silicon hydride $(SiH)_n$. Alcohol and chlorine reacted too, and the compound obtained had a composition $[SiH_{0.5}Cl_{0.33}(OR)_{0.17}]_n$. To resolve these difficulties, Schott[568] later reacted $CaSi_2$ with HBr in a $AlBr_3$ melt, and, depending on the duration of reaction and reaction temperature, obtained various layer compounds in which the fourth valency of the silicon is occupied by hydrogen or bromine, or is unoccupied, which gives it the character of a radical.

Kautsky and Haase[339, 340] demonstrated a new reaction with $SbCl_3$:

$$3\ CaSi_2 + 2\ SbCl_3 \longrightarrow 3\ CaCl_2 + 2\ Sb + 6\ Si.$$

This reaction yields a silicon compound with a layered polyradical. The Sb formed in the reaction can be washed out with an iodine solution, indicating that no reaction has taken place with the silicon. The brown silicon mixed with the resulting $CaCl_2$ (from which it cannot be washed out) is a very reactive solid, and spontaneously ignites with water or alcohol. Recent studies by Hengge *et al.* agree with these results. Bonitz[46], on the suggestion of Kautsky, continued these studies. The reaction of $CaSi_2$ with chlorine in an inert solvent yielded a layered silicon as described above, but an excess of chlorine produced polymeric chlorides and finally low-molecular-weight silicon chlorides[47].

Hengge and Scheffler[277] replaced the elemental chlorine by interhalide compounds, hoping that better reaction conditions might be attained by using a more polar reagent for attacking the polar Ca-Si linkage. $CaSi_2$ reacts explosively with iodine chloride in the absence of solvents. Milder conditions, with CCl_4 as a solvent, yield layered silicon monochloride:

$$CaSi_2 + 4\ ICl \longrightarrow CaCl_2 + \frac{2}{n}\ (SiCl)_n + 2\ I_2.$$

The free iodine can readily be washed out, but $CaCl_2$ can be removed only incompletely because all the solvents for this substance react with silicon chloride. The reaction will take place quantitatively only if the starting $CaSi_2$ has been finely ground under inert gas or in a solvent. Grinding in air leads to partial oxidation and the reaction remains incomplete. Silicon monochloride prepared in this way is a yellow scaly product whose analysis indicates the exact 1 : 1 silicon to chlorine ratio. The quantity of hydrogen liberated in alkaline medium is also consistent with the three Si-Si bonds per silicon atom required for the layer structure.

In silicon monochloride thus prepared the chlorine of the Si-Cl bond is available for a variety of reactions. Like all silicon-halogen bonds, Si-Cl reacts with alcohol to form alkoxy derivatives.

The reaction is substantially affected by steric factors. When an attempt is made in the molecular model to replace every chlorine by a methoxy group, a strong mutual hindrance between the methyl groups is observed. Complete replacement of chlorine is therefore impossible, as the experiment confirms. The reaction with methanol shows a good agreement with the model.

When longer-chain alcohols are used, the degree of replacement decreases in accordance with the increased space requirements. Other reactions (with amines, mercaptans, etc.), exhibit similar relationships.

The two-stage reaction of $CaSi_2$ with iodine chloride resulting in intermediate formation of a free silicon layer is demonstrated by the following example. We can react $CaSi_2$ with ICl in methanolic solution, when the methoxy product is obtained directly.

$$n\,CaSi_2 + 2n\,meOH + 2n\,ICl \longrightarrow n\,CaCl_2 + 2\,[Si(Ome)]_n + n\,I_2 + n\,H_2 \text{ found}$$

$$[Si(Ome)_{0.8}]_n.$$

The silicon valencies which cannot be occupied for steric reasons have a radical character and the Si : substituent ratio is less than one.

Heating with the less reactive IBr instead of ICl leads to quantitative formation of silicon monobromide as a light-brown powder:

$$n\,CaSi_2 + 4\,n\,IBr \longrightarrow n\,CaBr_2 + 2\,(SiBr)_n + 2\,n\,I_2.$$

However, under suitable mild conditions, the free silicon layer is obtained:

$$n\,CaSi_2 + 2\,n\,IBr \xrightarrow{0^\circ} n\,CaBr_2 + 2\,(Si)_n + n\,I_2.$$

This free silicon layer forms a dark-brown powder, which is stable under nitrogen, oxidizes rapidly in air with ignition, and reacts spontaneously with alcohol or water. It is identical with the scaly silicon, prepared by Kautsky (see p. 108).

The vacant silicon valencies of the silicon layer represent radicals, as was confirmed by the ESR spectra. The methoxy derivatives (with a deficiency of substituents) prepared from $CaSi_2$ plus ICl in methanolic solution likewise supply an ESR signal of an intensity which has the expected relationship to that of the layered silicon.

Careful fluorination[267)] of $(SiCl)_n$ or $(SiBr)_n$ layers with SbF_3 yields a layered silicon monofluoride $(SiF)_n$ in the form of a yellow, air-sensitive powder. The Si-F frequency in the IR spectra is extremely low ($725\ cm^{-1}$), indicating a very weak Si-F linkage.

$LiAlH_4$ reacts with $(SiBr)_n$ to a polymeric silicon hydride[267)] with the stoichiometric composition $(SiH)_n$, and $CaBr_2$. The compound is a brown powder which reacts with water evolving hydrogen, and burns spontaneously in air. The Si-H IR frequency is also extremely low, the value of $2100\ cm^{-1}$

being in good agreement with other compounds in which the silicon atom attached to the hydrogen is bonded to other silicon atoms[65, 265].

$CaSi_2$ reacts with ammonium halides to calcium halide and a silicon nitride with an Si : N ratio of 3 : 1[253]:

$$3n\ CaSi_2 + 6n\ NH_4Br \longrightarrow 2\ (Si_3N)_n + 4n\ NH_3 + 6n\ NH_3 + 3n\ CaBr_2.$$

Ammonium chloride and ammonium bromide are suitable for this reaction, while with the iodide and the fluoride there are other reactions. The reaction takes place with the finely powdered substances in the solid state in the absence of solvents. The reaction temperature is the sublimation temperature of the ammonium salts.

Most amines react with $CaSi_2$ in form of their hydrogen halides but the quarternary ammonium salts do not react[257]. The reaction becomes weaker with a larger organic group. The primary amines react better than the secondary and tertiary amines.

Two examples can be briefly given. The methylamine hydrobromide reacts with powdered $CaSi_2$ when heated at 300 °C in the absence of solvents:

$$3nCaSi_2 + 6nCH_3NH_2 \cdot HBr \longrightarrow [Si_6(NHCH_3)_{4.88}]_n + 1.12\ CH_3NH_2 + 5.44\ n\ H_2 + 3nCaBr_2.$$

The reaction products are the free amine, hydrogen, $CaBr_2$ and a brown powder. Here too, a compound with free radicals is indicated by the ESR spectrum. Since the methylamine group is nearly the same size as the methoxy group, the composition is analogous to the methoxy derivatives. The radical concentration has the same value.

A larger amine such as aniline hydrobromide would be expected to substitute less completely. Only two molecules can be placed on a model of an Si_6 ring, one above and one below the ring. This theoretical composition must have four radical positions on each ring. This composition was empirically confirmed, and so the equation can be written:

$$3nCaSi_2 + 6n\varphi NH_2 \cdot HBr \longrightarrow 3nCaSi_2 + [Si_6(NH\varphi)_2]_n + 4n\varphi NH_2 + 4nH_2.$$

The ESR spectrum also shows the expected size of the paramagnetic peak.

With phosphonium halides such as PH_4I, the phosphide Si_6P_2 is obtained:

$$3\ CaSi_2 \xrightarrow{6\ PH_4I} CaI_2 + (Si_6P_2)_n + 4\ PH_3 + 6\ H_2.$$

With sulphur compounds, similar sulphur-containing silicon layers are produced. For example, S_2Cl_2 reacts with $CaSi_2$ to form a solid compound $(SiS)_n$ in which the S atoms form S-S bridges on or between the layers[266]:

$$n\ CaSi_2 + n\ S_2Cl_2 \longrightarrow n\ CaCl_2 + 2\ (SiS)_n.$$

The acid hydrolysis and the determination of the resulting disulfide shows the presence of the S-S bridge. In this case, it was possible that the disulfide could have bridged two layers, but then the surface of the compound would have been very much smaller. Measurements showed the correct surface area so that a bridge between two layers is highly unlikely.

10.2.4. The Color of Polymeric Silicon Compounds

For siloxene (see p. 99) the color and the fluorescence of the derivatives are shifted toward longer wavelengths when the Si ring is subjected to the action of electrons from substituents carrying free electron pairs. This led to the postulation of an additional bond system within the Si_6 ring as the chromophoric part of the molecule.

Similar interactions should be important in the layer compounds $(SiX)_n$. Various derivatives are diamagnetic and some have radical character because of the size of their substituents. This group of substances does not exhibit fluorescence, so the spectra must be measured as the reemission spectra of the powders[268, 276, 556].

Table 8: Reemission spectra of $(SiX)_n$-derivatives

Substituent X	λ_{max} (nm)
F	263
Cl	313
CH_3	330
H	333
Br	360
S	376
OCH_3	385
$NHCH_3$	420

Table 8 gives the data for a wide variety of derivatives of SiX compounds. There is no clear shift of the absorption band, only a shifting and broadening of the longer-wavelength part of the band.

The electronegativities of the substituents are an important determinant of the shift, but the greater variety of the substances reveals a second effect due to the $d\pi$-$p\pi$ overlap between the substituents and the silicon. Studies of the origin of the color showed that the more electrons at the silicon, the greater the bond strength of the silicon-silicon bonds and the greater the color shift.

Thus, if we take the formula of Gordy to determine the electronegativity of the substituent and include the constants: bond order, force constant, and atomic distance with the usual values for bonds having $d\pi$-$p\pi$ character, we get in most cases a good general correspondence with the color. Determination of the electronegativity with the Gordy rule is not the best way to find the correlation between color and $d\pi$-$p\pi$ character or electronegativity, but it does indicate the rules valid for our compounds.

It is again plausible that an additional non-localized bond system develops within the Si layer, causing the color; $(SiX)_n$ derivatives exhibit a metallic lustre and a certain degree of conductivity, which supports this possibility. The dark color of the layered silicon and the relatively great stability of the radical state show that the effect is particularly strongly developed in this substance. The radical electrons presumably undergo substantial resonance interaction with the electron system of the Si-Si bonds, possibly giving a resonance-stabilized polyradical.

11. References

[1] Abedini, M., Dyke, C. H. van, MacDiarmid, A. G.: J. Inorg. Nucl. Chem. *25*, 307 (1963).
[2] Abedini, M., MacDiarmid, A. G.: Inorg. Chem. *2*, 608 (1963).
[3] Abedini, M., MacDiarmid, A. G.: Inorg. Chem. *5*, 2040 (1966).
[4] Abrahamson, E. W., Joffe, I., Post, H. W.: J. Org. Chem. *13*, 275 (1948).
[5] Anderson, H. H.: J. Am. Chem. Soc. *72*, 2761 (1950).
[6] Anderson, J. W., Drake, J. E.: J. Chem. Soc. *A*, 3131 (1970).
[7] Anderson, J. W., Drake, J. E.: J. Chem. Soc. *A*, 1424 (1971).
[8] Anderson, J. S., Ogden, J. S.: J. Chem. Phys. *51*, 4189 (1969).
[9] Andrejew, D. N.: J. Prakt. Chem. *23*, 288 (1964).
[10] Andrejew, D. N., Kucharskaja, E. W.: Angew. Chem. *36*, 2309 (1963).
[11] Andrews, T. D., Phillips, C. S. G.: J. Chem. Soc. *A*, 46 (1966).
[12] Andrianov, K. A., Khananashvili, L. M.: Organometal. Chem. Rev. *2*, 141 (1967).
[13] Antipin, P. F., Ssergejew, W. W.: Angew. Chem. *27*, 784 (1954).
[14] Amberger, E. Mühlhofer, E.: J. Organometal. Chem. *12*, 55 (1968).
[15] Armstrong, D. R., Perkins, P. G.: Theoret. Chim. Acta *5*, 69 (1966).
[16] Attridge, C. J.: J. Organometal. Chem. *13*, 259 (1968).
[17] Atwell, W. H., Mahone, L. G., Hayes, S. F., Uhlmann, J. H.: J. Organometal. Chem. *18*, 69 (1969).
[18] Atwell, W. H., Weyenberg, D. R.: Angew. Chem. *81*, 485 (1969).
[19] Atwell, W. H., Weyenberg, D. R.: J. Am. Chem. Soc. *90*, 3438 (1968).
[20] Atwell, W. H., Weyenberg, D. R.: J. Organometal. Chem. *5*, 594 (1966).
[21] Atwell, W. H., Weyenberg, D. R.: J. Organometal. Chem. *7*, 71 (1967).
[22] Baird, R. B., Sefcik, M. D., Ring, M. A.: Inorg. Chem. *10*, 883 (1971).
[23] Band, S. J., Davidson, I. M. T., Lambert, C. A., Stephanson, J. L.: Chem. Commun. *1967*, 723.
[24] Bartell, L. S., Clippard, F. B., Boates, T. L.: Inorg. Chem. *9*, 2436 (1970).
[25] Bassler, J. M., Timms, P. L., Margrave, J. L.: Inorg. Chem. *5*, 729 (1966).
[26] Beagley, B., Monaghan, J. J.: J. Mol. Structure: *8*, 401 (1971).
[27] Becher, H. J.: Fortschr. Chem. Forsch. *10*, 156 (1968).
[28] Beezer, A. E., Mortimer, C. T.: J. Chem. Soc. *A*, 514 (1966).
[29] Benkeser, R. A., Foster, D. J.: J. Am. Chem. Soc. *74*, 5314 (1952).
[30] Benkeser, R. A., Landesman, H., Foster, D. C.: J. Am. Chem. Soc. *74*, 648 (1952).
[31] Benkeser, B. A., Severson, R. G.: J. Am. Chem. Soc. *73*, 1424 (1951).
[32] Bentham, J. E., Cradock, S., Ebsworth, E. V.: Inorg. Nucl. Chem. Letters *7*, 1070 (1971).
[33] Besson, M.: Compt. Rend. *154*, 116a (1912).
[34] Besson, A., Fournier, L.: Compt. Rend. *149*, 34 (1909).
[35] Besson, A., Fournier, L.: Compt. Rend. *151*, 1055 (1910).
[36] Bethke, G. W., Wilson, M. K.: J. Chem. Phys. *26*, 1107 (1957).
[37] Bettler, C. R., Sendra, J. C., Urry, G.: Inorg. Chem. *9*, 1060 (1970).
[38] Billy, M.: Compt. Rend. *251*, 1639 (1960).
[39] Billy, M.: Compt. Rend. *250*, 4163 (1960).

40) Billy, M.: Bull. Soc. Chim. France *1961*, 1550.
41) Billy, M., Gomsat, P.: Compt. Rend. *269*, 919 (1969).
42) Bock, H., Enßlin, W.: Angew. Chem. *83*, 435 (1971).
43) Böhm, J., Hassel, O.: Z. Anorg. allgem. Chem. *160*, 152 (1927).
44) Bolduc, W. J., Ring, M. A.: J. Organometal. Chem. *6*, 202 (1966).
45) Bonhoeffer, K. F.: Z. Phys. Chem. *131*, 363 (1928).
46) Bonitz, E.: Chem. Ber. *94*, 220 (1961).
47) Bonitz, E.: Angew. Chem. *78*, 475 (1966).
48) Borer, K., Phillips, C. S. G.: Proc. Chem. Soc. *1959*, 189.
49) Bott, R. W., Eaborn, C., Rushton, B. M.: J. Organometal. Chem. *3*, 455 (1965).
50) Bourgeois, P., Calas, R., Duffaut, N.: Bull. Soc. Chim. France *1965*, 2694.
51) Bowery, M., Purnell, J. H.: J. Am. Chem. Soc. *92*, 2594 (1970).
52) Breederveld, H., Thoor, T. J. W. van, Waterman, H. I.: Res. Corresp. Suppl. to Research 7, No. 5, 29 (1954); Chem. Zentralblatt *54 S*, 1683 (1950/54).
53) Brockway, L. O., Beach, J. Y.: J. Am. Chem. Soc. *60*, 1836 (1938).
54) Brockway, L. O., Davidson, N. R.: J. Am. Chem. Soc. *63*, 3287 (1941).
55) Brook, A. G., Anderson, D. G., Duff, J. M., Jones, P. F., MacRae, D. M.: J. Am. Chem. Soc. *90*, 1076 (1968).
56) Brook, A. G., Gilman, H., Miller, L. S.: J. Am. Chem. Soc. *75*, 4759 (1953).
57) Brook, A. G., Mauris, R. J.: J. Am. Chem. Soc. *79*, 971 (1957).
58) Brown, M. P., Fowles, G. W. A.: J. Chem. Soc. *1958*, 2811.
59) Brown, M. P., Webster, D. E.: J. Phys. Chem. *64*, 698 (1960).
60) Brune, H. A., Schulte, D.: Chem. Ber. *100*, 3438 (1967).
61) Bürger, H.: Fortschr. Chem. Forsch. *9*, 1 (1967).
62) Bürger, H., Eujen, R.: Z. Anorg. Allgem. Chem. in press.
63) Bürger, H., Falius, H.: Z. Anorg. Allgem. Chem. *363*, 24 (1968).
64) Bürger, H., Goetze, U.: Angew. Chem. *80*, 192 (1968).
65) Bürger, H., Kilian, W.: J. Organometal. Chem. *26*, 47 (1971).
66) Bürger, H., Kilian, W.: J. Organometal. Chem. *18*, 299 (1969).
67) Bürger, H., Kilian, W., Burczyk, K.: J. Organometal. Chem. *21*, 291 (1970).
67a) Bürger, H., Eujen, R.: Low-valent silicon. In: Topics Curr. Chem. *50*, 1 (1974).
68) Burkhard, C. A.: J. Am. Chem. Soc. *71*, 963 (1949).
69) Bygden, A.: Ber. Deut. Chem. Ges. *45*, 707 (1912).
70) Calas, R., Bourgeois, P.: Bull. Soc. Chim. France *1968*, 4678.
71) Calas, R., Frainnet, E., Dentone, Y.: Compt. Rend. *259*, 3777 (1964).
72) Calas, R., Marchand, A., Frainnet, E., Gervall, P.: Bull. Soc. Chim. France *1968*, 2478.
73) Carberry, E. Dombek, B. D.: J. Organometal. Chem. *22*, C 43 (1970).
74) Carberry, E., West, R.: J. Organometal. Chem. *6*, 582 (1966).
75) Carberry, E., West, R.: J. Am. Chem. Soc. *91*, 5440 (1969).
76) Carberry, E., West, R., Glass, G. E.: J. Am. Chem. Soc. *91*, 5446 (1969).
77) Cartledge, F. K.: J. Organometal. Chem. *13*, 516 (1968).
78) Cartledge, F. K., Riedel, K. H.: J. Organometal. Chem. *34*, 11 (1972).
79) Cerato, C. C., Lauer, J. L., Beachell, H. C.: J. Chem. Phys. *22*, 1 (1954).
80) Cetini, G., Castiglioni, M., Volpe, P., Gambrina, O.: Ric. Sci. *39*, 392 (1969); C. A. *73*, 20 515 g (1970).
81) Chainini, G. R., Cooper, S., Gilman, H.: J. Org. Chem. *28*, 2146 (1963).
82) Chambers, D. B., Glockling, F.: J. Chem. Soc. *A*, 735 (1968).
83) Chang, C. H., Porter, R. F., Bauer, S. H.: J. Phys. Chem. *74*, 1363 (1970).
84) Chernyshev, E. A., Komalenkova, N. G., Bashkirova, S. A.: Zh. Obshch. Khim. *41*, 1175 (1971).
85) Chernyshev, E. A., Tostikova, N. G., Ioffe, S. L., Petrow, A. D.: Zh. Obshch. Khim. *32*, 369 (1962).
86) Chernyshev, E. A., Tolstikova, N. G., Ivashenko, A. A, Zelenetskaya, A. A., Leiters, L. A.: Izv. Akad. Nauk SSSR 660 (1963); C. A. 7553 g (1963).
87) Chernyshev, E. A., Tolstikova, N. G., Ivashenko, A. A., Zelenetskaya, A. A., Leiters, L. A.: Izv. Akad. Nauk SSSR *4*, 666 (1963).

88) Ch ..rd, R. P., Gowenlock, B. G., Johnson, A. F., Stevenson, J.: J. Organometal. Chem. *3 H*, 53 (1972).
89) Connolly, J. W., Stocklon, Ben F., Emerick, C. M., Bernheim, Bernheim, E. A.: Inorg. Chem. *9*, 93 (1970).
90) Connor, J. A., Finney, G., Leigh, G. J., Haszeldine, R. N., Robinson, P. J., Sedgwick, R. D., Simmons, R. F.: Chem. Commun. *1966*, 178.
91) Connor, J. A., Haszeldine, R. N., Leigh, G. J., Sedgwick, R. D.: J. Chem. Soc. *A*, 768 (1967).
92) Cooper, G. D., Gilbert, A. R.: J. Am. Chem. Soc. *82*, 5042 (1960).
93) Coulin, R. T., Gaspar, P. P., Levin, R. H., Maitland, J.: J. Am. Chem. Soc. *94*, 7165 (1972).
94) Cox, A. P., Varma, R.: J. Chem. Phys. *44*, 2619 (1966).
95) Craig, A. D., MacDiarmid, A. G.: J. Inorg. Nucl. Chem. *24*, 161 (1962).
96) Craig, A. D., Urenovitch, J. V., MacDiarmid, A. G.: J. Chem. Soc. *1962*, 548.
97) Cremer, E., Faessler, A., Krämer, H.: Naturwissenschaften *46*, 377 (1954).
98) Cuvigny, Th., Normant, H.: J. Organometal. Chem. *38*, 217 (1972).
99) Daniel, S. H., Gennaro, G. P., Ranck, K. M., Tang, Y. N.: J. Phys. Chem. *76*, 1249 (1972).
100) Davidson, I. M. T.: J. Organometal. Chem. *24*, 97 (1970).
101) Davidson, I. M. T., Stephanson, J. L.: J. Chem. Soc. *A*, 282 (1968).
102) Davidson, I. M. T., Stephenson, I. L.: Chem. Commun. *1966*, 746.
103) Drake, J. E., Goddard, N.: Inorg. Nucl. Chem. Letters *4*, 385 (1968).
104) Drake, J. E., Goddard, N.: J. Chem. Soc. *A*, 662 (1969).
105) Drake, J. E., Goddard, N.: J. Chem. Soc. *A*, 2587 (1970).
106) Drake, J. E., Goddard, N., Riddle, C.: J. Chem. Soc. *A*, 2704 (1969).
107) Drake, J. E., Goddard, N., Simpson, J.: Inorg. Nucl. Chem. Letters *4*, 361 (1968).
108) Drake, J. E., Goddard, N., Westwood, N. P. C.: J. Chem. Soc. *A*, 3305 (1971).
109) Drake, J. E., Jolly, W. L.: Chem. Ind. (London) 1470 (1962).
110) Drake, J. E., Jolly, W. L.: J. Chem. Phys. *38*, 1033 (1963).
111) Drake, J. E. Simpson, J.: Inorg. Nucl. Chem. Letters *2*, 219 (1966).
112) Drake, J. E., Westwood, N. P. C.: Chem. Ind. (London) *1969*, 24.
113) Drake, J. E., Westwood, N. P. C.: J. Chem. Soc. *A*, 2587 (1971).
114) Drake, J. E., Westwood, N. P. C.: J. Chem. Soc. *A*, 3300 (1971).
115) Dubois, I., Herzberg, G., Varma, R. D.: J. Chem. Phys. *47*, 4262 (1967).
116) Duffaut, N., Dunogues, J., Calas, R.: Compt. Rend. *C 268*, 967 (1969).
117) Duncan, J. L.: Spectrochim. Acta *20*, 1807 (1964).
118) Dyke, C. H. van, MacDiarmid, A. G.: Inorg. Chem. *3*, 747 (1964).
119) Dyke, C. H. van, MacDiarmid, A. G.: Inorg. Chem. *3*, 1071 (1964).
120) Dyke, C. H. van, MacDiarmid, A. G.: J. Inorg. Nucl. Chem. *25*, 1503 (1963).
121) Eaborn, C., Jackson, R. A., Walsingham, R. W.: Chem. Commun. 300 (1965).
122) Ebsworth, E. A. V.: Volatile Silicon Compounds, Pergamon Press, London *1963*, 85.
123) Ebsworth, E. A. V., Turner, J. J.: Trans. Faraday Soc. *60*, 256 (1964).
124) Egorov, Yu. P., Kirei, G. G., Samvilenko, S. A., Chernyshev, E. A., Tolstikova, N. G.: Izv. Akad. Nauk SSSR *1963*, 569.
125) Emeleus, H. J., Reid, C.: J. Chem. Soc. *1939*, 1021.
126) Emeleus, H. J., Stewart, K.: J. Chem. Soc. *1936*, 677.
127) Emeleus, H. J., Tufail, M.: J. Inorg. Nucl. Chem. *29*, 2081 (1967).
128) Emons, H. H., Boenicke, H.: J. Prakt. Chem. *18*, 11 (1962).
129) Emons, H. H., Hellmold, P.: Z. Anorg. Allgem. Chem. *341*, 78 (1965).
130) Emons, H. H., Hellmold, P.: Z. Anorg. Allgem. Chem. *355*, 265 (1967).
131) Emons, H. H., Theisen, L.: Monatsh. Chem. *103*, 62 (1972).
132) Estacio, P., Sefcik, M. D., Chan, E. K., Ring, M. A.: Inorg. Chem. *9*, 1068 (1970).
133) Faessler, A.: Angew. Chem. *84*, 51 (1972).
134) Fearon, F. W. G., Gilman, H.: J. Organometal. Chem. *9*, 403 (1967).
135) Feher, F., Fischer, H.: Naturwissenschaften *51*, 461 (1964).

136) Feher, F., Kuhlbörsch, G., Luhleich, H.: Z. Anorg. Allgem. Chem. *303*, 294 (1960).
137) Feher, F., Kuhlbörsch, G., Luhleich, M.: Z. Anorg. Allgem. Chem. *303*, 283, (1960).
138) Feher, F., Kuhlbörsch, G., Luhleich, H.: Z. Naturforsch. *146*, 466 (1959).
139) Feher, F., Kuhlbörsch, G., Luhleich, M., Keller, H.: Angew. Chem. *70*, 402 (1958).
140) Feher, F., Jansen, G., Rohmer, H.: Z. Anorg. Allgem. Chem. *329*, 31 (1964).
141) Feher, F., Plichta, P., Guillery, R.: Chem. Ber. *103*, 3028 (1970).
142) Feher, F., Plichta, P., Guillery, R.: Tetrahedron Letters *1970*, 2889.
143) Feher, F., Plichta, P., Guillery, R.: Inorg. Chem. *10*, 606 (1971).
144) Feher, F., Schinkitz, D., Schaaf, J.: Z. Anorg. Allgem. Chem. *383*, 303 (1971).
145) Feher, F., Schinkitz, D., Strack, H.: Z. Anorg. Allgem. Chem. *385*, 202 (1971).
146) Feher, F., Schinkitz, D., Wronka, G.: Z. Anorg. Allgem. Chem. *384*, 226 (1971).
147) Feher, F., Strack, H.: Naturwissenschaften *50*, 570 (1963).
148) Feher, F., Tromm, W.: Z. Anorg. Allgem. Chem. *282*, 29 (1955).
149) Finholt, A. E., Bond, A. C., Wilzbach, K. E., Schlesinger, H. I.: J. Am. Chem. Soc. *69*, 2692 (1947).
150) Fontal, B., Spiro, T.: Inorg. Chem. *10*, 9 (1971).
151) Frainnet, E., Calas, R., Gerval, P.: Compt. Rend. *261*, 1329 (1965).
152) Friedel, C.: Ber. Deut. Chem. Ges. *2*, 57 (1869).
153) Friedel, C.: Compt. Rend. *73*, 1011 (1871).
154) Friedel, C., Ladenburg, A.: Ann. Chim. Phys. *19*, 390 (1880).
155) Friedel, C., Ladenburg, A.: Liebigs Ann. Chem. *203*, 241 (1880).
156) Fritz, G.: Z. Naturforsch. *5 b*, 444 (1950).
156a) Fritz, G.: Organometallic synthesis of carbosilanes. In: Topics Curr. Chem. *50*, 43 (1974).
157) Fritz, G., Kummer, H.: Z. Anorg. Allgem. Chem. *306*, 191 (1960).
158) Gaidis, J. M., Briggs, P. R., Shannon, T. W.: J. Phys. Chem. *75*, 974 (1971).
159) Gaj, B. J., Gilman, H.: Chem. Ind. (London) *1960*, 493.
160) Gamboa, J. M.: An. Soc. esp. Fis. Quim: *46 B*, 699 (1950); C. A. *49*, 6766 (1955).
161) Garrity, S. P., Ring, M. A.: Inorg. Nucl. Chem. Lett. *4*, 77 (1968).
162) Gaspar, P. P, Bock, S. A., Eckelmann, W. C.: J. Am. Chem. Soc. *90*, 6914 (1968).
163) Gaspar, P. P., Levy, C. A., Adair, G. M.: Inorg. Chem. *9*, 1272 (1970).
164) Gaspar, P. P., Markusch, P.: Chem. Commun. *1970*, 1331.
165) Gaspar, P., Markusch, P., Holten, J. D., Frost, J. J.: J. Phys. Chem. *76*, 1352 (1972).
166) Gattermann, L., Weinling, K.: Ber. Deut. Chem. Ges. *27*, 1943 (1894).
167) Gattermann, L., Ellery, E.: Ber. Deut. Chem. Ges. *32*, 1114 (1899).
168) Geisler, T. C., Cooper, C. G., Norman, A. D.: Inorg. Chem. *11*, 1710 (1972).
169) Geller, S., Thurmond, C. D.: J. Am. Chem. Soc. *77*, 5285 (1955).
170) George, M. V., Bernard, J. G., Gilman, H.: J. Org. Chem. *24*, 624 (1959).
171) George, M. V., Lichtenwalter, G. D., Gilman, H.: J. Am. Chem. Soc. *81*, 978 (1959).
172) Gilman, H., Lichtenwalter, G. D., Wittenberg, D.: J. Am. Chem. Soc. *81*, 5320 (1959).
173) George, M. V., Peterson, D. J., Gilman, H.: J. Am. Chem. Soc. *82*, 403 (1960).
174) Gerval, P., Calas, R., Frainnet, E., Dunogues, J.: J. Organometal. Chem. *31*, 329 (1971).
175) Gilman, H., Aoki, D.: J. Org. Chem. *24*, 426 (1959).
176) Gilman, H., Aoki, D.: Chem. Ind. (London) *1960*, 1165.
177) Gilman, H., Aoki, D.: J. Organometal. Chem. *1*, 449 (1964).
178) Gilman, H., Atwell, W. H.: J. Org. Chem. *28*, 2905 (1963).
179) Gilman, H., Atwell, W. H.: J. Organometal. Chem. *4*, 176 (1965).
180) Gilman, H., Atwell, W. H., Cartledge, F. K: Advan. Organometal. Chem. *4*, 1 (1966).
181) Gilman, H., Atwell, W. H., Schwebke, G. L.: J. Organometal. Chem. *2*, 369 (1964).
182) Gilman, H., Atwell, W. H., Schwebke, G. L.: Chem. Ind. (London) *1964*, 1063.
183) Gilman, H., Atwell, W. H., Sen, P. K., Smith, C. L.: J. Organometal. Chem. *4*, 163 (1965).
184) Gilman, H., Chapman, D. R.: Chem. Ind. (London) *1965*, 1788.
185) Gilman, H., Chapman, D. R.: J. Organometal. Chem. *5*, 392 (1966).
186) Gilman, H., Chapman, D. R.: J. Organometal. Chem. *8*, 451 (1967).
187) Gilman, H., Chapman, D. R., Schwebke, G. L.: J. Organometal. Chem. *14*, 267 (1968).

188) Gilman, H., Cottis, S. G., Atwell, W. H.: J. Am. Chem. Soc. *1964*, 1596.
189) Gilman, H., Dunn, G. E.: J. Am. Chem. Soc. *73*, 5077 (1951).
190) Gilman, H., Dunn, G. E.: Chem. Rev. *52*, 94 (1953).
191) Gilman, H., Goodman, J. J.: J. Am. Chem. Soc. *75*, 1250 (1953).
192) Gilman, H., Gorsich, R. D.: J. Am. Chem. Soc. *80*, 3243 (1958).
193) Gilman, H., Harrell, R. L.: J. Organometal. Chem. *5*, 199 (1966).
194) Gilman, H., Harrell, R. L: J. Organometal. Chem. *9*, 67 (1967).
195) Gilman, H., Harrell, R., Chang, K. J., Cottis, S.: J. Organometal. Chem. *2*, 434 (1964).
196) Gilman, H., Holmes, J. M., Smith, C. L.: Chem. Ind. (London) *1965*, 848.
197) Gilman, H., Ingham, R. K., Smith, A. G.: J. Org. Chem. *18*, 1743 (1953).
198) Gilman, H., Inoue, S.: J. Org. Chem. *29*, 3418 (1964).
199) Gilman, H., Inoue, S.: Chem. Ind. (London) *1964*, 74.
200) Gilman, H., Jarvie, A. W. P.: Chem. Ind. (London) *1960*, 965.
201) Gilman, H., Jarvie, A. W. P.: Chem. Ind. (London) *1960*, 1271.
202) Gilman, H., Lichtenwalter, G. D.: J. Am. Chem. Soc. *80*, 608 (1958).
203) Gilman, H., Lichtenwalter, G. D.: J. Org. Chem. *24*, 1588 (1959).
204) Gilman, H., Morris, P. S.: J. Organometal. Chem. *6*, 102 (1966).
205) Gilman, H., Petersen, D. J., Jarvie, A. W. J., Winkler, H. J. S.: J. Am. Chem. Soc. *82*, 2076 (1960).
206) Gilman, H., Petersen, D. J., Jarvie, A. W. P., Winkler, H. J. S.: Tetrahedron Letters *23*, 5 (1960).
207) Gilman, H., Peterson, D. J., Tomasi, R. A., Harrell, R. L.: J. Organometal. Chem. *4*, 167 (1965).
208) Gilman, H., Peterson, D. J., Wittenberg, D.: Chem. Ind. (London) *1958*, 1479.
209) Gilman, H., Schwebke, G. L.: J. Am. Chem. Soc. *85*, 1016 (1963).
210) Gilman, H., Schwebke, G. L.: J. Am. Chem. Soc. *86*, 2693 (1964).
211) Gilman, H., Schwebke, G. L.: Advan. Organometal. Chem. *1*, 89 (1964).
212) Gilman, H., Schwebke, G. L.: J. Organometal. Chem. *3*, 382 (1965).
213) Gilman, H., Shiina, K.: J. Organometal. Chem. *8*, 369 (1967).
214) Gilman, H., Shiina, K., Aoki, D., Gay, B. J., Wittenberg, D., Brennan, T.: J. Organometal. Chem. *13*, 323 (1968).
215) Gilman, H., Smart, R.: J. Org. Chem. *15*, 720 (1950).
216) Gilman, H., Smith, C. L.: J. Am. Chem. Soc. *86*, 1454 (1964).
217) Gilman, H., Smith, C. L.: J. Organometal. Chem. *6*, 665 (1966).
218) Gilman, H., Smith, C. L.: J. Organometal. Chem. *8*, 245 (1967).
219) Gilman, H., Smith, C. L.: J. Organometal. Chem. *14*, 91 (1968).
220) Gilman, H., Tomasi, R. A.: J. Org. Chem. *27*, 3647 (1962).
221) Gilman, H., Tomasi, R. A.: J. Org. Chem. *28*, 1651 (1963).
222) Gilman, H., Tomasi, R. A.: Chem. Ind. (London) *1963*, 954.
223) Gilman, H., Tomasi, R. A.: J. Organometal. Chem. *9*, 223 (1967).
224) Gilman, H., Wu, T. C.: J. Am. Chem. Soc. *73*, 4031 (1951).
225) Gilman, H., Wu, T. C.: J. Am. Chem. Soc. *75*, 3762 (1953).
226) Gilman, H., Wu, T. C.: J. Org. Chem. *25*, 2251 (1960).
227) Gilman, H., Wu, T. C., Hartzfeld, H. A., Guter, G. A., Smith, A. G., Goodman, J. J., Eidt, S. H.: J. Am. Chem. Soc. *74*, 561 (1952).
228) Gmelin, Handbuch der anorg. Chemie, *15* (Silicium) part B.
229) Gohlke, R. S.: J. Am. Chem. Soc. *90*, 2713 (1968).
230) Gokhale, S. D., Drake, S. E., Jolly, W. L.: J. Inorg. Nucl. Chem. *27*, 1911 (1965).
231) Gokhale, S. D., Jolly, W. L.: Inorg. Chem. *3*, 946 (1964).
232) Gokhale, S. D., Jolly, W. L.: Inorg. Chem. *3*, 1141 (1964).
233) Gokhale, S. D., Jolly, W. L.: Inorg. Chem. *4*, 596 (1965).
234) Groschwitz, E. A., Ingle, W. M., Ring, M. A.: J. Organometal. Chem. *9*, 421 (1967).
235) Grube, G., Speidel, H.: Z. Elektrochem. *53*, 341 (1949).
236) Gunn, S. R., Green, L. G.: J. Phys. Chem. *65*, 779 (1961).
237) Gunn, S. R., Green, L. G.: J. Phys. Chem. *68*, 946 (1964).
238) Gutowsky, H. S., Ripmeester, J. A.: J. Chem. Phys. *56*, 1332 (1972).

239) Gutowsky, H. S., Stejskal, E. O.: J. Chem. Phys. *22*, 939 (1954).
240) Hague, D. N., Prince, R. H.: Proc. Chem. Soc. *1962*, 300.
241) Hague, D. N., Prince, R. H.: Chem. Ind. (London) *1964*, 1492.
242) Hague, D. N., Prince, R. H.: J. Chem. Soc. *1965*, 4690.
243) Hamada, K., Robinson, E. A.: Bull. Chem. Soc. Japan *45*, 2219 (1972).
244) Harada, Y., Murell, J. N., Sheena, H. H.: Chem. Phys. Lett. *1*, 595 (1968).
245) Hayashi, M.: J. Chem. Soc. Japan *79*, 775 (1958).
246) Harrah, L. A., Ryan, M. T., Tamborski, C.: Spectrochim. Acta *18*, 21 (1962).
247) Hastie, J. W., Hauge, R. H., Margrave, J. L.: Inorg. Chim. Acta 3, 601 (1969).
248) Hayashi, M.: J. Chem. Soc. Japan *78*, 1472 (1957); C. A. *52*, 5129 (1958).
249) Hellner, E.: Z. Anorg. Allgem. Chem. *261*, 226 (1950).
250) Hengge, E.: Monatsh. Chem. *102*, 734 (1971).
251) Hengge, E.: Halogen-Chemistry, Vol. II, p. 169ff. London: Academic Press 1967.
252) Hengge, E.: Chem. Ber. *95*, 645 (1962).
253) Hengge, E.: J. Anorg. Allgem. Chem. *315*, 298 (1962).
254) Hengge, E., Abushaban, M.: Allg. Prakt. Chem. *18*, 393 (1967).
255) Hengge, E., Bauer, G.: Angew. Chem. *85*, 304 (1973).
256) Hengge, E., Brychcy, U.: Monatsh. Chem. *97*, 1309 (1966).
257) Hengge, E., Brychcy, U.: Z. Anorg. Allgem. Chem. *339*, 120(1965).
258) Hengge, E., Grupe: Chem. Ber. *97*, 1783 (1964).
259) Hengge, E., Holtschmidt, N.: J. Organometal. Chem. *12*, P 5 (1968).
260) Hengge, E., Holtschmidt, N.: Monatsh. Chem. *99*, 340 (1968).
261) Hengge, E., Innertsberger, E.: Diplomarbeit Graz 1967.
262) Hengge, E., Kollmann, G.: unpublished.
263) Hengge, E., Lunzer, F.: Syn. Inorg. Metal-org. Chem. *2*, 93 (1972).
264) Hengge, E., Marketz, H.: Monatsh. Chem. *100*, 890 (1969).
265) Hengge, E., Marketz, H.: Monatsh. Chem. *101*, 528 (1970).
266) Hengge, E., Olbrich, G.: Z. Anorg. allgem. Chem. *365*, 321 (1969).
267) Hengge, E., Olbrich, G.: Monatsh. Chem. *101*, 1068 (1970).
268) Hengge, E., Olbrich, G.: Monatsh. Chem. *101*, 1074 (1970).
269) Hengge, E., Olbrich, G., Georgiadon, E.: Monatsh. Chem. *99*, 686 (1968).
270) Hengge, E., Pletka, H. D., Höfler, F.: Monatsh. Chem. *101*, 325 (1970).
271) Hengge, E.: Chem. Ber. *95*, 648 (1962).
272) Hengge, E., Pretzer, K.: Chem. Ber. *96*, 470 (1963).
273) Hengge, E., Reuter, H.: Naturwissenschaften *49*, 514 (1962).
274) Hengge, E., Reuter, H., Petzold, R.: Z. Naturforsch. *18 b*, 425 (1963).
275) Hengge, E., Scheffler, G.: Naturwissenschaften *50*, 474 (1963).
276) Hengge, E., Scheffler, G.: Monatsh. Chem. *95*, 1461 (1964).
277) Hengge, E., Scheffler, G.: Monatsh. Chem. *95*, 1450 (1964).
278) Hengge, E., Starz, E., Strubert, W.: Monatsh. Chem. *99*, 1787 (1968).
279) Hengge, E., Strubert, W.: Z. Anorg. allgem. Chem. *389*, 83 (1972).
280) Hengge, E. Wolfer, D.: Angew. Chem. *85*, 304 (1973).
281) Hengge, E., Wolfer, D.: J. Organometal. Chem., *66*,413 (1974).
282) Hertwig, K. A.: Z. Naturforsch. *6b*, 337 (1951).
283) Hertwig, K. A., Wiberg, E.: Z. Naturforsch. *6b* , 336 (1951).
284) Hess, G. G., Lampe, F. W., Sommer, L. H.: J. Am. Chem. Soc. *86*, 3174 (1964).
285) Hess, G. G., Lampe, F. W., Sommer, L. H.: J. Am. Chem. Soc. *87*, 5327 (1965).
286) Higuchi, T., Shimada, A.: Bull. Chem. Soc. Japan *39*, 1316 (1966).
287) Higuchi, T., Shimada, A.: Bull. Chem. Soc. Japan *40*, 752 (1967).
288) Hoch, M., Johnston, H. L.: J. Am. Chem. Soc. *75*, 5224 (1953).
289) Höfler, F.: unpublished.
290) Höfler, F., Hengge, E.: Monatsh. Chem. *103*, 1506 (1972).
291) Höfler, F., Hengge, E.: Monatsh. Chem. *103*, 1513 (1972).
292) Höfler, F., Sawodny, W., Hengge, E.: Spectrochim. Acta *26A*, 819 (1970).
293) Höfler, F., Waldhör, S., Hengge, E.: Spectrochim. Acta *28A*, 29 (1972).

294) Höfler, F., Wolfer, D.: unpublished.
295) Hollandsworth, R. P., Ingle, W. M., Ring, M. A.: Inorg. Chem. *6*, 844 (1967).
296) Hollandsworth, R. P., Ring, M. A.: Inorg. Chem. *7*, 1635 (1968).
297) Holm, J. W. C., Chantry, A. W., Borrow, R. F.: Trans. Faraday Soc. *54*, 1580 (1958).
298) Hopkins, H. P., Thompson, J. C., Margrave, J. L.: J. Am. Chem. Soc. *90*, 901 (1968).
299) Hosomi, A., Sakurai, H.: Bull. Chem. Soc. Japan *45*, 248 (1972).
300) Hosomi, A., Sakurai, H.: J. Am. Chem. Soc. *94*, 1384 (1972).
301) Husk, G. R., West, R.: J. Am. Chem. Soc. *87*, 3993 (1965).
302) Husk, G. R., Wexler, R., Kilcullen, B. M.: J. Organometal. Chem. *29*, C 49 (1971).
303) Indriksons, A., West, R.: J. Am. Chem. Soc. *92*, 6704 (1970).
304) Ingle, W. M., Groschwitz, E. A., Ring, M. A.: Inorg. Chem. *6*, 1429 (1967).
305) Ipatieff, V. N., Dolgov, B. N.: Ber. Deut. Chem. Ges. *62B*, 1220 (1929).
306) Iseard, B. S., Pedley, J. B., Treverton, J. A.: J. Chem. Soc. *A*, 3095 (1971).
307) Ishikawa, M., Kumada, M.: Chem. Commun. *1969*, 567.
308) Ishikawa, M., Kumada, M.: Chem. Commun. *1970*, 157.
309) Ishikawa, M., Kumada, M.: Chem. Commun. *1970*, 612.
310) Ishikawa, M., Kumada, M.: Syn. Inorg. Metal-org. Chem. *1*, 191 (1971).
311) Ishikawa, M., Kumada, M.: Syn. Inorg. Metal-org. Chem. *1*, 229 (1971).
312) Ishikawa, M., Kumada, M.: Chem. Commun. *1971*, 489.
313) Ishikawa, M., Kumada, M.: J. Organometal. Chem. *42*, 325 (1972).
314) Ishikawa, M., Kumada, M, Sakurai, H.: J. Organometal. Chem. *23*, 63 (1970).
315) Ishikawa, M., Takaoka, T., Kumada, M.: J. Organometal. Chem. *42*, 333 (1972).
316) Janzon, K., Schäfer, H., Weiss, A.: Z. Naturforsch. *23b*, 1544 (1968).
317) Jarvie, A. W. P., Gilman, H.: J. Org. Chem. *26*, 1999 (1961).
318) Jarvie, A. W. P., Winkler, H. S., Gilman, H.: J. Org. Chem. *27*, 614 (1962).
319) Jarvie, A. W. P., Winkler, H. S., Peterson, D. J., Gilman, H.: J. Am. Chem. Soc. *83*, 1921 (1961).
320) Johannesen, R. B., Farrar, T. C., Brinckman, F. E., Coyle, T. D.: J. Chem. Phys. *44*, 962 (1966).
321) Johnson, F., Gohlke, R. S., Nasutavicus, W. A.: J. Organometal. Chem. *3*, 233 (1965).
322) Johnson, W. C., Hogness, T. R.: J. Am. Chem. Soc. *56*, 1252 (1934).
323) Johnson, W. C., Isenberg, S.: J. Am. Chem. Soc. *57*, 1349 (1935).
324) Kaczmarczyk, A., Millard, M., Nuss, J. W., Urry, G.: J. Inorg. Nucl. Chem. *26*, 421 (1964).
325) Kaczmarczyk, A., Millard, M., Urry, G.: J. Inorg. Nucl. Chem. *17*, 186 (1961).
326) Kaczmarczyk, A., Nuss, J. W., Urry, G.: J. Inorg. Nucl. Chem. *26*, 427 (1964).
327) Kaczmarczyk, A., Urry, G.: J. Am. Chem. Soc. *82*, 751 (1960).
328) Kaczmarczyk, A., Urry, G.: J. Inorg. Nucl. Chem. *26*, 415 (1964).
329) Kana'an, A. S., Margrave, J. L.: Inorg. Chem. *3*, 1037 (1964).
330) Kanazashi, M., Takakura, M.: Bull. Chem. Soc. Japan *35*, 1840 (1962).
331) Kartsev, G. N., Kokoreva, I., Syrkin, Ya. K., Mironov, V. F., Chernyshev, E. A. Z. strukturnoj. Chim. *6*, 309 (1965); C. A. *63*, 5071c (1965).
332) Katayama, M., Shimanouchi, P., Morino, Y., Mizushima, S.: J. Chem. Soc. Japan *70*, 19 (1949).
333) Katayama, M., Shimanouchi, T., Morino, Y., Mizushima, S.: J. Chem. Phys. *18*, 506 (1950).
334) Kautsky, H.: unpublished.
335) Kautsky, H.: Z. Anorg. Allgem. Chem. *117*, 209 (1921).
336) Kautsky, H.: Z. Naturforsch. *7b*, 174 (1952).
337) Kautsky jr, H.: Kolloid.-Z. *153*, 109 (1957).
338) Kautsky, H., Fritz, G., Richter, H.: Z. Naturforsch. *9b*, 236 (1954).
339) Kautsky, H., Haase, L.: Chem. Ber. *86*, 1226 (1953).
340) Kautsky, H., Haase, L.: Z. Naturforsch. *8b*, 45 (1953).
341) Kautsky, H., Herzberg, G.: Z. Anorg. Allgem. Chem. *139*, 135 (1924).
342) Kautsky, H., Herzberg, G.: Chem. Ber. *1924*, 1665.
343) Kautsky, H., Hirsch, A.: Z. Anorg. Allgem. Chem. *170*, 1 (1928).

344) Kautsky, H., Kautsky jr, H.: Chem. Ber. *89*, 571 (1956).
345) Kautsky, H., Keck, H., Kunze, H.: Z. Naturforsch. *9b*, 165 (1954).
346) Kautsky, H., Neitzke, O.: Z. Physik *31*, 60 (1925).
347) Kautsky, H., Pfleger, H.: Z. Anorg. Allgem. Chem. *295*, 206 (1958).
348) Kautsky, H., Richter, T.: Z. Naturforsch. *11b*, 365 (1956).
349) Kautsky, H., Siebel, H. P.: Z. Anorg. Allgem. Chem. *273*, 113 (1953).
350) Kautsky, H., Thiele, H.: Z. Anorg. Allgem. Chem. *144*, 197 (1925).
351) Kautsky, H., Thiele, H.: Z. Anorg. Allgem. Chem. *173*, 115 (1928).
352) Kautsky, H., Vogell, W., Oeters, F.: Z. Naturforsch. *10b*, 597 (1955).
353) Kautsky, H., Zocher, H.: Z. Physik *9*, 267 (1922).
354) Köster, A.: Angew. Chem. *69*, 563 (1957).
355) Kumada, M., Ishikawa, M.: J. Organometal. Chem. *1*, 153 (1963).
356) Kumada, M., Ishikawa, M.: J. Organometal. Chem. *1*, 411 (1964).
357) Kumada, M., Ishikawa, M.: J. Organometal. Chem. *6*, 451 (1966).
358) Kumada, M., Ishikawa, M., Maeda, S.: J. Organometal. Chem. *2*, 478 (1964).
359) Kumada, M., Ishikawa, M., Maeda, S.: J. Organometal. Chem. *5*, 120 (1966).
360) Kumada, M., Ishikawa, M., Murai, B.: Kogyo Kagaku Zasshi *66*, 637 (1963); C. A. 15303 (1963).
361) Kumada, M., Ishikawa, M., Sakamoto, S., Maeda, S.: J. Organometal. Chem. *17*, 223 (1969).
362) Kumada, M., Ishikawa, M., Tamao, K.: J. Organometal. Chem. *5*, 226 (1966).
363) Kumada, M., Maeda, S.: Inorg. Chim. Acta *1*, 105 (1967).
364) Kumada, M., Naka, K., Ishikawa, M.: J. Organometal. Chem. *2*, 136 (1964).
365) Kumada, M., Nakajima, J., Ishikawa, M., Yamamoto, Y.: J. Org. Chem. *23*, 292 (1958).
366) Kumada, M., Sakumoto, S., Ishikawa, M.: J. Organometal. Chem. *17*, 231 (1969).
367) Kumada, M., Shiina, K., Yamaguchi, M.: J. Chem. Soc. Japan *57*, 230 (1954); C. A. *49*, 11542c (1955).
368) Kumada, M., Tamao, K.: Advan. Organometal. Chem. *6*, 19 (1968).
369) Kumada, M., Tamao, K., Ishikawa, M., Matsuno, M.: Chem. Commun. *1968*, 614.
370) Kumada, M., Tamao, K., Tabuko, T., Ishikawa, M.: J. Organometal. Chem. *9*, 43 (1967).
371) Kumada, M., Yamaguchi, M., Yamamoto, Y., Nakajima, J., Shiina, K.: J. Org. Chem. *21*, 1264 (1956).
372) Kummer, D., Köster, H.: Angew. Chem. *81*, 897 (1969).
373) Kummer, D., Köster, H.: Angew. Chem. *83*, 408 (1971).
374) Kummer, D., Köster, H., Speck, M.: Angew. Chem. *81*, 574 (1969).
375) Kelling, H.: Z. Chem. *7*, 237 (1967).
376) Kelling, H.: Z. Chem. *8*, 391 (1968).
377) Kennedy, R. C., Freeman, L. P., Fox, A. P., Ring, M. A.: J. Inorg. Nucl. Chem. *28*, 1373 (1966).
378) Kempter, C. P., Alvarez-Tostado, C.: Z. Anorg. Allgem. Chem. *290*, 238 (1957).
379) Khanna, V. M., Besenbruch, G., Margrave, J. L.: J. Chem. Phys. *46*, 2310 (1967).
380) Khanna, V. M. Hauge, R., Curl, R. F., Margrave, J. L.: J. Chem. Phys. *47*, 5031 (1967).
381) Kipping, F. S.: J. Chem. Soc. *119*, 647 (1921).
382) Kipping, F. S.: J. Chem. Soc. *123*, 2590 (1923).
383) Kipping, F. S.: J. Chem. Soc. *123*, 2598 (1923).
384) Kipping, F. S.: J. Chem. Soc. *1924*, 2291,
385) Kipping, F. S.: J. Chem. Soc. *1927*, 2728.
386) Kipping, F. S., Sands, J. E.: J. Chem. Soc. *119*, 830 (1921).
387) Kipping, F. S., Sands, J. E.: J. Chem. Soc. *119*, 848 (1921).
388) Kipping, F. S., Thompson, R. A.: J. Chem. Soc. *1928*, 1989.
389) Kohanek, J. J., Estacio, P., Ring, M. A.: Inorg. Chem. *8*, 2516 (1969).
390) Kowalew, I. F.: Optik u. Spektroskopie *3*, 552 (1957).
391) Kraus, C. A., Nelson, W. K.: J. Am. Chem. Soc. *56*, 195 (1934).
392) Krüerke, U.: Chem. Ber. *95*, 177 (1962).

393) Kühlein, K., Neumann, W. P.: J. Organometal. Chem. *14,* 317 (1968).
394) Lappert, M. F., Pedley, J. B., Simpson, J., Spalding, T. R.: J. Organometal. Chem. *29,* 195 (1971).
395) Lebeau, P., Le Chatelier, M.: Compt. Rend. *148,* 43 (1909).
396) Lieser, K. H., Kohlschütter, H. W.: Z. Anorg. Allgem. Chem. *313,* 199 (1961).
397) Liu, C. S., Margrave, J. L., Thompson, J. C., Timms, P. L.: Can. J. Chem. *50,* 459 (1972).
398) Liu, C. S., Margrave, J. L., Thompson, J. C.: Can. J. Chem. *50,* 465 (1972).
399) Liu, C. S., Thompson, J. C.: Inorg. Chem. *10,* 1100 (1971).
400) Liu, C. S., Thompson, J. C.: J. Organometal. Chem. *38,* 249 (1972).
401) Lowe, J. P., Parr, R. G.: J. Chem. Phys. *44,* 3001 (1966).
402) MacDiarmid, A.: J. Inorg. Nucl. Chem. *25,* 1534 (1963).
403) MacDiarmid, A., Rabel, E.: unpublished.
404) MacDiarmid, A., Urenovitsch, J. V.: J. Chem. Soc. *1963,* 1091.
405) MacKay, K. K., Hosfield, S. T., Stobart, S. R.: J. Chem. Soc. *A,* 2937 (1969).
406) MacKay, K. K., Watt, R.: Organometal. Chem. Rev. *4,* 137 (1969).
407) McLean, A. D., Yoshimina, M.: Tables of linear molecule wavefunction, IBM Journal Research Development Supplement.
408) Maddock, A. G., Emeleus, H. J.: J. Chem. Soc. *1946,* 1131.
409) Margrave, J. L., Kanaan, A. S., Pease, D. C.: J. Phys. Chem. *66,* 1200 (1962).
410) Margrave, J. L., Sharp, K. G., Wilson, P. W.: J. Am. Chem. Soc. *92,* 1530 (1970).
411) Margrave, J. L., Sharp, K. G., Wilson, P. W.: J. Inorg. Nucl. Chem. *32,* 1817 (1970).
412) Margrave, J. L., Williams, D. L., Wilson, P. W.: Inorg. Nucl. Chem. Letters 7, 103 (1971).
413) Martin, G.: Ber. Deut. Chem. Ges. *46,* 2442 (1913).
414) Martin, G.: Ber. Deut. Chem. Ges. *46,* 3289 (1913).
415) Martin, G.: J. Chem. Soc. *105,* 2836 (1914).
416) Martin, G.: J. Chem. Soc. *105,* 2860 (1914).
417) Martin, G.: J. Chem. Soc. *107,* 319 (1915).
418) Martin, G.: J. Chem. Soc. *107,* 1043 (1915).
419) Mustapha M'hirsi, Brini, M.: Bull. Soc. Chim. France *1968,* 1509.
420) Moissan, H.: Bull. Soc. Chim. France *29,* 443 (1903).
421) Moissan, H., Smiles, S.: Bull. Soc. Chim. France *27,* 1190 (1902).
422) Moissan, H., Smiles, S.: Compt. Rend. *134,* 569 (1902).
423) Morino, Y.: J. Chem. Phys. *24,* 164 (1956).
424) Morino, Y., Hirota, E.: J. Chem. Phys. *28,* 185 (1958).
425) Morris, P. J., Gilman, H.: J. Organometal. Chem. *11,* 463 (1968).
426) Morrison, J. A., Ring, M. A.: Inorg. Chem. *6,* 100 (1967).
427) Murato, H., Shimizu, K.: J. Chem. Phys. *23,* 1968 (1955).
428) Nakadaira, Y., Sakurai, H.: J. Organometal Chem. *47,* 61 (1973).
429) Naumann, K., Zon, G., Mislow, K.: J. Am. Chem. Soc. *91,* 2788 (1969).
430) Nay, M. A., Woodall, G. N. C., Strausz, O. P., Gunning, H. E.: J. Am. Chem. Soc. *87,* 179 (1965).
431) Nebergall, W. H., Johnson, O. H.: J. Am. Chem. Soc. *71,* 4022 (1949).
432) Nefedov, O. M., Manakov, M. N.: Angew. Chem. *76,* 270 (1964).
433) Nefedov, O. M., Manakov, M. N.: Angew. Chem. *78,* 1039 (1966).
434) Neumann, W. P., Kühlein, K.: Tetrahedron Letters *23,* 1541 (1963).
435) Normann, A. D.: Chem. Commun. *1968,* 812.
436) Nuss, J. W., Urry, G.: J. Inorg. Nucl. Chem. *26,* 435 (1964).
437) Ogier, J.: Ann. Chim. *20,* 32 (1880).
438) Okawara, R., Tanaka, T.: Bull. Chem. Soc. Japan *27,* 119 (1954).
439) Okawara, R., Tanaka, T., Maruo, K.: Bull. Chem. Soc. Japan *28,* 189 (1955).
440) Owen, P. W., Skell, P. S.: Tetrahedron Letters *1972,* 1807.
441) Palmer, K. W., Kipping, F. S.: J. Chem. Soc. *1930,* 1020.
442) Parrish, W.: Acta Cryst. *13,* 838 (1960).
443) Petro, A. J., Smyth, C. P.: J. Am. Chem. Soc. *79,* 6147 (1957).
444) Pfeiffer, M., Spangenberg, H. J.: Z. Physik. Chem. (Leipzig) *232,* 47 (1966).

445) Pflugmacher, A., Rohrmann, J.: Z. Anorg. Allgem. Chem. *290,* 101 (1957).
446) Phillips, C. S. G., Powell, P., Semlyen, J. A., Timms, P. L.: Z. Anal. Chem. *197,* 202 (1963).
447) Phillips, C. S. G., Timms, P. L.: J. Chromatog. *5,* 131 (1961).
448) Piekoś, R., Radechi, A.: Z. Anorg. Allgem. Chem. *309,* 258 (1961).
449) Pike, R. M., Moynahan, E. B.: Inorg. Chem. *6,* 168 (1967).
450) Pitt, C. G.: J. Am. Chem. Soc. *91,* 6613 (1969).
451) Pitt, C. G.: J. Am. Chem. Soc. *92,* 519 (1970).
452) Pitt, C. G., Bock, H.: Chem. Commun. *1972,* 28.
453) Pitt, C. G., Carey, R. N., Toren, E. C.: J. Am. Chem. Soc. *94,* 3806 (1972).
454) Pitt, C. G., Fowler, M. S.: J. Am. Chem. Soc. *89,* 6792 (1967).
455) Pitt, C. G., Fowler, M. S.: J. Am. Chem. Soc. *90,* 1928 (1968).
456) Pitt, C. G., Habercorn, M. S., Bursey, M. M., Rogerson, P. F.: J. Organometal. Chem. *15,* 359 (1968).
457) Pitt, C. G., Jones, L. L., Ramsey, B. G.: J. Am. Chem. Soc. *89,* 5471 (1967).
458) Pitzer, K. S.: J. Am. Chem. Soc. *70,* 2140 (1948).
459) Potter, H. N.: Trans. Am. Electrochem. Soc. *12,* 191, (1907).
460) Purnell, J. H., Walsh, R.: Proc. Roy. Soc. *A 293,* 543 (1961).
461) Quig, J. B., Wilkinson, J. A.: J. Am. Chem. Soc. *48,* 902 (1926).
462) Rao, V. M., Curl jr., R. F.: J. Chem. Phys. *45,* 2032 (1966).
463) Rao, V. M., Curl, R. F., Timms, P. L., Margrave, J. L.: J. Chem. Phys. *43,* 2557 (1965).
464) Rao, V. M., Venkateswa, P.: J. Mol. Spectroscopy *7,* 287 (1961).
465) Raymonda, J. W., Muenter, J. S., Klemperer, W. A.: J. Chem. Phys. *52,* 3458 (1970).
466) Razuvaev, G. A., Vyazankin, N. S., D'yachkovskaya, O. S., Kiseleva, J. G., Dergunov, Yu. I.: Zh. Obshch. Khim. *31,* 4056 (1961); C. A. *1962,* 9875b.
467) Reedy, D., Urry, G.: Inorg. Chem. *6,* 2117 (1967).
468) Ring, M. A., Baird, R. R., Estacio, P.: Inorg. Chem. *9,* 1004 (1970).
469) Ring, M. A., Freemann, L. P., Fox, A. P.: Inorg. Chem. *3,* 1200 (1964).
470) Ring, M. A., Pneutes, M. J., O'Neal, H. E.: J. Am. Chem. Soc. *92,* 4845 (1970).
471) Ring, M. A., Ritter, D. M.: J. Am. Chem. Soc. *83,* 802 (1961).
472) Roark, D. N., Peddle, G. J. D.: J. Am. Chem. Soc. *94,* 5837 (1972).
473) Robinson, E. A.: Can. J. Chem. *41,* 3021 (1963).
474) Rochow, E. G., Didtschenko, R.: J. Am. Chem. Soc. *74,* 5545 (1952).
475) Royen, P., Rocktäschel, C.: Angew. Chem. *76,* 302 (1964).
476) Royen, P., Rocktäschel, C.: Z. Anorg. Allgem. Chem. *346,* 279 (1966).
477) Ryan, J. W.: J. Am. Chem. Soc. *84,* 4730 (1962).
478) Saalfeld, F. E., Svec, H. J.: Inorg. Chem. *2,* 50 (1963).
479) Saalfeld, F. E., Svec, H. J.: Inorg. Chem. *3,* 1442 (1964).
480) Saalfeld, F. E., Svec, H. J.: J. Phys. Chem. *70,* 1753 (1966).
481) Sakurai, H., Deguchi, S., Yamagata, M., Morimoto, S., Kira, M., Kumada, M.: J. Organometal. Chem. *18,* 285 (1969).
482) Sakurai, H., Hosomi, A.: J. Am. Chem. Soc. *93,* 1709 (1971).
483) Sakurai, H., Hosomi, A.: J. Organometal. Chem. *36,* C 15 (1972).
484) Sakurai, H., Hosomi, A., Kumada, M.: Chem. Commun. *1968,* 930.
485) Sakurai, H., Hosomi, A., Kumada, M.: Chem. Commun. *1969,* 1969.
486) Sakurai, H., Hosomi, A., Kumada, M.: Bull. Chem. Soc. Japan *44,* 568 (1971).
487) Sakurai, H., Hosomi, A., Nakajima, J., Kumada, M.: Bull. Chem. Soc. Japan *39,* 2263 (1966).
488) Sakurai, H., Imoto, T., Hayashi, N., Kumada, M.: J. Am. Chem. Soc. *87,* 4001 (1965).
489) Sakurai, H., Kira, M., Kumada, M.: Bull. Chem. Soc. Japan *41,* 1494 (1968).
490) Sakurai, H., Kira, M., Kumada, M.: Bull. Chem. Soc. Japan *44,* 1167 (1971).
491) Sakurai, H., Kira, M., Sato, T.: J. Organometal. Chem. *42,* C 24 (1972).
492) Sakurai, H., Kishida, T., Hosomi, A., Kumada, M.: J. Organometal. Chem. *8,* 65 (1967).
493) Sakurai, H., Kobayashi, Y., Nakadaira, Y.: J. Am. Chem. Soc. *93,* 5272 (1971).
494) Sakurai, H., Koh, R., Hosomi, A., Kumada, M.: Bull. Chem. Soc. Japan *39,* 2050 (1966).

495) Sakurai, H., Kumada, M.: Bull. Chem. Soc. Japan *37*, 1894 (1964).
496) Sakurai, H., Okada, A.: J. Organometal. Chem. *35*, C 13 (1972).
497) Sakurai, H., Okada, A.: J. Organometal. Chem. *36*, C 13 (1972).
498) Sakurai, H., Okada, A., Kira, M., Yonezawa, K.: Tetrahedron Letters *19*, 1511 (1971).
499) Sakurai, H., Tominga, K., Kumada, M.: Bull. Chem. Soc. Japan *39*, 1279 (1966).
500) Sakurai, H., Tominga, K., Kumada, M.: Bull. Chem. Soc. Japan *39*, 1820 (1966).
501) Sakurai, H., Tominga, K., Watanabe, T.: Tetrahedron Letters *1966*, 5493.
502) Sakurai, H., Yamamori, H., Kumada, M.: Chem. Commun. *1968*, 198.
503) Sakurai, H., Yamamori, H., Kumada, M.: Bull. Chem. Soc. Japan *38*, 2024 (1965).
504) Sakurai, H., Yamamori, H., Kumada, M.: J. Org. Chem. *33*, 1527 (1968).
505) Sakurai, H., Yamamori, H., Yamageta, M., Kumada, M.: Bull. Chem. Soc. Japan *41*, 1474 (1968).
506) Sakurai, H., Watanabe, T., Kumada, M.: J. Organometal. Chem. *7*, P 15 (1967).
507) Sakurai, H., Watanabe, T., Kumada, M.: J. Organometal. Chem. *9*, P 11 (1967).
508) Saper, P. G.: Phys. Rev. *42*, 498 (1932).
509) Selin, T. G., West, R.: Tetrahedron *5*, 97 (1959).
510) Sharp, K. G., Margrave, J. L.: Inorg. Chem. *8*, 2655 (1969).
511) Shaw, C. F., Allred, A. L.: J. Organometal. Chem. *28*, 53 (1971).
512) Shaw, C. F., Allred, A. L.: Inorg. Chem. *10*, 1340 (1971).
513) Shiina, K., Kumada, M.: J. Org. Chem. *23*, 139 (1958).
514) Shiina, K., Kumada, M.: J. Chem. Soc. Japan *60*, 1395 (1957); C. A. *53*, 17889 (1959).
515) Shiina, K., Kumada, M.: Mem. Fac. Eng. Osaka City Univ. *2*, 128 (1960); C. A. *55*, 25732 (1961).
516) Shiina, K., Minoura, Y.: unpublished, ref. in 386.
517) Shimizu, K., Murato, H.: J. Chem. Soc. Japan *77*, 343 (1956).
518) Skell, P. S., Goldstein, E. J.: J. Am. Chem. Soc. *86*, 1442 (1964).
519) Skell, P. S., Owen, P. W.: J. Am. Chem. Soc. *89*, 3933 (1967).
520) Skell, P. S., Owen, P. W.: J. Am. Chem. Soc. *94*, 5434 (1972).
521) Skinner, H. A., Pichler, G.: Quart. Rev. *17*, 264 (1963).
522) Smith, D. L., Kirk, R., Timms, P. L.: Chem. Commun. *1972*, 295.
523) Smyth, C. P.: US Gort. Res. Repts. *32*, 115 (1959); C. A. *1960*, 16961 c.
524) Solan, D., Burg, A. B.: Inorg. Chem. *11*, 1253 (1972).
525) Solan, D., Timms, P. L.: Inorg. Chem. *7*, 2157 (1968).
526) Sommer, L. H.: Stereochemistry, Mechanismn and Silicon. New York: McGraw Hill Book Co. Inc. 1965.
527) Sommer, L. H., Mason, R.: J. Am. Chem. Soc. *87*, 1619 (1965).
528) Sommer, L. H., Rosborough, K. T.: J. Am. Chem. Soc. *89*, 1756 (1967).
529) Sommer, L . H., Rosborough, K. T.: J. Am. Chem. Soc. *91*, 7067 (1969).
530) Sommer, L. H., Rosborough, K. T., McLich, J.: J. Am. Chem. Soc. *94*, 4217 (1972).
531) Spanier, E. J., MacDiarmid, A. G.: Inorg. Chem. *1*, 432 (1962).
532) Spialter, L., Austin, J. D.: J. Am. Chem. Soc. *88*, 1828 (1966).
533) Suga, H., Saki, S.: Bull. Chem. Soc. Japan *32*, 1088 (1959).
534) Suga, H., Saki, S., Chatani, Y., Taguchi, J.: Ann. Inst. Fiber Res. *8*, 51 (1954).
535) Sundermeyer, W.: Z. Anorg. Allgem. Chem. *310*, 50 (1961).
536) Swick, D. A., Karle, J. L.: J. Chem. Phys. *23*, 1499 (1955).
537) Székely, T., Nefedov, O. M., Garzo, G., Shiryayer, V. I., Fritz, D.: Acta Chim. Acad. Sci. Hung. *54*, 241 (1967).
538) Schaaf, T. F., Oliver, J. P.: J. Am. Chem. Soc. *91*, 4327 (1969).
539) Schäfer, H.: Z. Anorg. Allgem. Chem. *274*, 265 (1953).
540) Schäfer, H.: Chem. transportreaktion. Weinheim: Verlag Chemie 1962.
541) Schäfer, H., Hörnle, R.: Z. Anorg. Allgem. Chem. *263*, 261 (1950).
542) Schäfer, H., Morcher, B.: Z. Anorg. Allgem. Chem. *290*, 279 (1957).
543) Schäfer, H., Nickl, J.: Z. Anorg. Allgem. Chem. *274*, 250 (1953).
544) Schaschel, E. T., Gray, D. N., Timms, P. L.: J. Organometal. Chem. *35*, 69 (1972).
545) Schenk, P. W., Bloching, H.: Z. Anorg. Allgem. Chem. *334*, 57 (1964).
546) Schlenk, W. J., Renning, J., Racky, G.: Ber. Deut. Chem. Ges. *44*, 1178 (1911).

547) Schmeisser, M.: Angew. Chem. *66,* 713 (1954).
548) Schmeisser, M.: Silicium-Schwefel-Phosphor. In: JUPAC-Colloquium Münster/Westf. Weinheim: Verlag Chemie 1954.
549) Schmeisser, M., Ehlers, K. P.: Angew. Chem. *76,* 781 (1964).
550) Schmeisser, M., Friedrich, K.: Angew. Chem. *76,* 782 (1964).
551) Schmeisser, M., Schwarzmann, M.: Z. Naturforsch. *11b*, 278 (1956).
552) Schmeisser, M., Voss, P.: Z. Anorg. Allgem. Chem. *334,* 50 (1964).
553) Schmeisser, M., Voss, P.: Fortschr. Chem. Forsch. *9,* 165 (1967).
554) Schmidbaur, H.: Angew. Chem. *82,* 773 (1970).
555) Scholl, R. L., Maciel, G. E., Musker, W. K.: J. Am. Chem. Soc. *94,* 6376 (1972).
556) Schott, G.: Z. Chem. *3,* 41 (1963).
557) Schott, G.: Fortschr. Chem. Forsch. *9,* 60 (1967).
558) Schott, G., Bünger, H. G.: Z. Anorg. Allgem. Chem. *362,* 71 (1968).
559) Schott, G., Gostmeier, G.: Z. Anorg. Allgem. Chem. *312*, 322 (1961).
560) Schott, G., Harzdorf, C.: Z. Anorg. Allgem. Chem. *306,* 180 (1960).
561) Schott, G., Herrmann, W.: Z. Anorg. Allgem. Chem. *288*, 1 (1956).
562) Schott, G., Herrmann, E.: Z. Anorg. Allgem. Chem. *307,* 97 (1960).
563) Schott, G., Herrmann, W., Hirschmann, R.: Angew. Chem. *68,* 213 (1956).
564) Schott, G., Hirschmann, E.: Z. Anorg. Allgem. Chem. *288,* 9 (1956).
565) Schott, G., Langecker, W.: Z. Anorg. Allgem. Chem. *358,* 210 (1968).
566) Schott, G., Meyer, J.: Z. Anorg. Allgem. Chem. *313,* 107 (1961).
567) Schott, G., Naumann, D.: Z. Anorg. Allgem. Chem. *291,* 103 (1957).
568) Schott, G., Naumann, D.: Z. Anorg. Allgem. Chem. *291,* 112 (1957).
569) Schulze, D.: Chem. Techn. *8,* 341 (1956).
570) Schumb, W. C.: Inorg. Synth. *2,* 98 (1946).
571) Schumb, W. C., Ackermann jr., J., Saffer jr., G. M.: J. Am. Chem. Soc. *60,* 2486 (1938).
572) Schumb, W. C., Gamble, E. L.: J. Am. Chem. Soc. *53,* 3191 (1931).
573) Schumb, W. C., Gamble, E. L.: J. Am. Chem. Soc. *54,* 583 (1932).
574) Schumb, W. C., Gamble, E. L.: Inorg. Synth. *1,* 42 (1939).
575) Schumb, W. C., Lefever, R. A.: J. Am. Chem. Soc. *75,* 1513 (1953).
576) Schumb, W. C., Lefever, R. A.: J. Am. Chem. Soc. *76,* 2091 (1954).
577) Schumb, W. C., Saffer jr., C. M.: J. Am. Chem. Soc. *61*, 363 (1939).
578) Schumb, W. C., Saffer jr., C. M.: J. Am. Chem. Soc. *63,* 93 (1941).
579) Schwarz, R.: Angew. Chem. *51,* 328 (1938).
580) Schwarz, R., Danders, Ch.: Chem. Ber. *80,* 444 (1947).
581) Schwarz, R., Danders, Ch.: Z. Anorg. Allgem. Chem. *253,* 273 (1947).
582) Schwarz, R., Gregor, V.: Z. Anorg. Allgem. Chem. *241,* 395 (1939).
583) Schwarz, R., Heinrich, F.: Z. Anorg. Allgem. Chem. *221,* 277 (1935).
584) Schwarz, R., Köster, A.: Z. Anorg. Allgem. Chem. *270,* 2 (1952).
585) Schwarz, R., Köster, A.: Z. Naturforsch. *7b,* 57 (1952).
586) Schwarz, R., Meckbach, H.: Z. Anorg. Allgem. Chem. *232,* 241 (1937).
587) Schwarz, R., Pflugmacher, A.: Ber. Deut. Chem. Ges. *75,* 1062 (1942).
588) Schwarz, R., Pietsch, G.: Z. Anorg. Allgem. Chem. *232,* 249 (1937).
589) Schwarz, W., Sexauer, W.: Ber. Deut. Chem. Ges. *59B,* 333 (1926).
590) Schwarz, R., Thiel, R.: Z. Anorg. Allgem. Chem. *235,* 247 (1938).
591) Steele, A. R., Kipping, F. S.: J. Chem. Soc. *1928,* 1431.
592) Steele, A. R., Kipping, F. S.: J. Chem. Soc. *1929,* 2545.
593) Steel, W. C., Stone, F. G. A.: J. Am. Chem. Soc. *84,* 3599 (1962).
594) Steudel, W., Gilmann, H.: J. Am. Chem. Soc. *82,* 6129 (1960).
595) Steward, O. W., Uhl, W. L., Sands, B. W.: J. Organometal. Chem. *15,* 329 (1968).
596) Stitt, F., Yost, D. M.: J. Chem. Phys. *5,* 90 (1937).
597) Stock, A., Brandt, A., Fischer, H.: Ber. Deut. Chem. Ges. *58,* 643 (1925).
598) Stock, A., Somieski, K.: Ber. Deut. Chem. Ges. *49,* 111 (1916).
599) Stock, A., Somieski, K.: Ber. Deut. Chem. Ges. *53,* 759 (1920).
600) Stock, A., Somieski, K.: Ber. Deut. Chem. Ges. *54,* 524 (1921).

601) Stock, A., Somieski, C.: Ber. Deut. Chem. Ges. *55*, 3961 (1922).
602) Stock, A., Somieski, C.: Ber. Deut. Chem. Ges. *56*, 247 (1923).
603) Wintgen, R.: Ber. Deut. Chem. Ges. *52*, 724 (1919).
604) Stock, A., Stiebeler, P.: Ber. Deut. Chem. Ges. *56*, 1087 (1923).
605) Stock, A., Stiebeler, P., Zeidler, F.: Ber. Deut. Chem. Ges. *56*, 1695 (1923).
606) Stokland, K.: Trans. Faraday Soc. *44*, 545 (1948).
607) Stolberg, U. G.: Angew. Chem. *74*, 696 (1962).
608) Stolberg, U. G.: Angew. Chem. *75*, 206 (1963).
609) Stolberg, U. G.: Z. Naturforsch. *18b*, 765 (1963).
610) Stolberg, U. G.: Chem. Ber. *96*, 2798 (1963).
611) Stolberg, U. G., Fritz, H. P.: Z. Anorg. Allgem. Chem. *330*, 1 (1964).
612) Takahashi, U.: Bull. Chem. Soc. Japan *28*, 443 (1955).
613) Takano, T., Kasai, N., Kakudo, M.: Bull. Chem. Soc. Japan *36*, 585 (1963).
614) Taketa, A., Kumada, M., Tarama, K.: J. Chem. Soc. Japan *60*, 201 (1957).
615) Taketa, A., Kumada, M., Tarama, K.: J. Chem. Soc. Japan, ind. Chem. Sect. *60*, 1392 (1957).
616) Taketa, A., Kumada, M., Tarama, K.: J. Chem. Soc. Japan *78*, 999 (1957).
617) Tamao, K., Ishikawa, M., Kumada, M.: Chem. Commun. *1969*, 73.
618) Tamao, K., Ishikawa, M., Kumada, M.: Chem. Commun. *1969*, 74.
619) Tamao, K., Kumada, M.: J. Organometal. Chem. *30*, 329 (1971).
620) Tamao, K., Kumada, M.: J. Organometal. Chem. *30*, 339 (1971).
621) Tamao, K., Kumada, M.: J. Organometal. Chem. *30*, 349 (1971).
622) Tamao, K., Kumada, M.: J. Organometal. Chem. *31*, 35 (1971).
623) Tamao, K., Kumada, M., Ishikawa, M.: J. Organometal. Chem. *31*, 17 (1971).
624) Tamao, K., Kumada, M., Noro, A.: J. Organometal. Chem. *31*, 169 (1971).
625) Tamao, K., Kumada, M., Sugimoto, T.: Chem. Commun. *1970*, 285.
626) Tang, Y. N., Gennaro, G. P., Su, Y. Y.: J. Am. Chem. Soc. *94*, 4355 (1972).
627) Tebben, E. M., Ring, M. A.: Inorg. Chem. *8*, 1787 (1969).
628) Thompson, H. W.: Spectrochim. Acta *16*, 238 (1960).
629) Thompson, J. C., Margrave, J. L.: Science *155*, 669 (1967).
630) Thompson, J. C., Margrave, J. L.: Inorg. Chem. *11*, 913 (1972).
631) Thompson, J. C., Margrave, J. L., Timms, P. L.: Chem. Commun. *16*, 566 (1966).
632) Timms, P. L., Ehlert, T. C., Margrave, J. L., Brinckmann, F. E., Farrar, T. C., Coyle, T. D.: J. Am. Chem. Soc. *87*, 3819 (1965).
633) Timms, P. L., Kent, R. A., Ehlert, T. C., Margrave, J. L.: Nature *207*, 187 (1965).
634) Timms, P. L., Kent, R. A., Ehlert, T. C., Margrave, J. L.: J. Am. Chem. Soc. *87*, 2824 (1965).
635) Timms, P. L., Phillips, C. S. G.: Inorg. Chem. *3*, 606 (1964).
636) Timms, P. L., Simpson, C. C., Phillips, C. S. G.: J. Chem. Soc. *1964*, 1467.
637) Timms, P. L., Stump, D. D., Kent, R. A., Margrave, J. L.: J. Am. Chem. Soc. *88*, 940 (1966).
638) Toral, T.: An. Soc. espan. Fisica Quim. *23*, 225 (1935); Chem. Zentr. *I*, 300 (1936).
639) Troost, L., Hautefeuilie, P.: Bull. Soc. Chim. France *16*, 240 (1871).
640) Troost, L., Hautefuilie, P.: Ann. Chim. Phys. 7, 459 (1876).
641) Tsai, T. T., Lehn, W. L., Marshall jr., C. J.: J. Org. Chem. *31*, 3047 (1966).
642) Urenovitsch, J. V., MacDiarmid, A. G., Nixon, E. R.: Appl. Spectr. *19*, 80 (1965).
643) Urenovitsch, J. V., MacDiarmid, A. G.: J. Am. Chem. Soc. *85*, 3372 (1963).
644) Urenovitsch, J. U., West, R.: J. Organometal. Chem. *3*, 138 (1965).
645) Urry, G.: Acc. Chem. Res. *3*, 306 (1970).
646) Urry, G.: J. Inorg. Nucl. Chem. *26*, 409 (1964).
647) Urry, G., Kaczmarcyk, A.: J. Am. Chem. Soc. *82*, 751 (1960).
648) Vanderwielen, A. J., Ring, M. A.: Inorg. Chem. *11*, 246 (1972).
649) Volpin, M. E., Koreshkov, Yu. D., Dulova, V. G., Kusanov, D. N.: Tetrahedron *18*, 107 (1962).
650) Voronkov, M. G., Khudobin, Yu. I.: J. Anorg. Allgem. Chem. *26*, 584 (1956); C. A. *50*, 13729 (1956).

651) Vrandi-Piskou, D., Parissakis, G.: J. Chromatog. *22,* 449 (1966).
652) Vyazankin, N. S., Razuvaev, G. A., Gladyshev, E. N., Gurikova, T. G.: Dokl. Akad. Nauk SSSR *155,* 1108 (1964); C. A. *61,* 1885 (1964 II).
653) Vyazankin, N. S., Razuvaev, G. A., Gladyshev, E. N., Korneva, S. P.: J. Organometal. Chem. *7,* 353 (1967).
654) Wang, Chi-Tao., Chou, H. Ch., Hung, M. S.: Hua Hsuch Pao *30,* 91 (1964); C. A. *61,* 1888 (1964).
655) Wannagat, U.: Angew. Chem. *76,* 234 (1964).
656) Wannagat, U., Brandstätter, O.: Angew. Chem. *75,* 345 (1963).
657) Wannagat, U., Brandstätter, O.: Monatsh. Chem. *94,* 1090 (1963).
658) Wannagat, U., Brandstätter, O.: Monatsh. Chem. *97,* 1352 (1966).
659) Wannagat, U., Hensen, K., Vielberg, F.: Monatsh. Chem. *99,* 431 (1968).
660) Wannagat, U., Meier, S.: Z. Anorg. Allgem. Chem. *392,* 179 (1972).
661) Wannagat, U., Schreiner, G., Brandstätter, O., Peach, M.: Monatsh. Chem. *96,* 1902 (1965).
662) Ward, L. G. L., MacDiarmid, A. G.: J. Am. Chem. Soc. *82,* 2151 (1960).
663) Ward, L. G. L., MacDiarmid, A. G.: J. Inorg. Nucl. Chem. *20,* 345 (1961).
664) Ward, L. G. L., MacDiarmid, A. G.: J. Inorg. Nucl. Chem. *21,* 287 (1962).
665) Wartenberg von, H.: Z. Elektrochem. *53,* 343 (1949).
666) Weidenbruch, M., Abrotat, G., John, K.: Chem. Ber. *104,* 2124 (1971).
667) West, R., Bailey, R. E.: J. Am. Chem. Soc. *85,* 287 (1963).
668) West, R., Indriksons, A.: J. Am. Chem. Soc. *94,* 6110 (1972).
669) West, R., Kramer, F. A., Carberry, E., Kumada, M., Ishikawa, N.: J. Organometal. Chem. *8,* 79 (1967).
670) White, D. G., Rochow, E. G.: J. Am. Chem. Soc. *76,* 3897 (1954).
671) Whitemore, F. C., Sommer, L. H., Gold. J.: J. Am. Chem. Soc. *69,* 1976 (1947).
672) Wiberg, E., Neumaier, A.: Angew. Chem. *74,* 514 (1962).
673) Wiberg, E., Stecher, O., Andrascheck, H. J., Kreuzbichler, L., Staude, E.: Angew. Chem. *75,* 516 (1963).
674) Wiberg, E., Stecher, O., Neumaier, A.: Inorg. Nucl. Chem. Letters *1,* 31 (1965).
675) Wiberg, E., Stecher, O., Neumaier, A.: Inorg. Nucl. Chem. Letters *1,* 33 (1965).
676) Wieland, K., Heise, M.: Angew. Chem. *63,* 438 (1951).
677) Wilde, R. E.: J. Mol. Spectr. *8,* 455 (1962).
678) Wilkins, C. J.: J. Chem. Soc. *1953,* 3409.
679) Wilson, G. R., Smith, A. G.: J. Org. Chem. *26,* 557 (1961).
680) Winkler, H. J. S., Gilman, H.: J. Org. Chem. *26,* 1265 (1961).
681) Winkler, H. J. S., Gilman, H.: J. Org. Chem. *27,* 254 (1962).
682) Winkler, H. J. S., Jarvie, A. W. P., Peterson, D. J., Gilman, H.: J. Am. Chem. Soc. *83,* 4089 (1961).
683) Wirtgen, R.: Ber. Deut. Chem. Ges. *52B,* 724 (1919).
684) Wittenberg, D., Gilman, H.: Quart. Rev. (London) *13,* 116 (1959).
685) Wittenberg, D., George, M. V., Gilman, H.: J. Am. Chem. Soc. *81,* 4812 (1959).
686) Vyazankin, N. S., Razuvaev, G. A., Kruglaya, O. A.: Izv. Akad. Nauk SSSR *1962,* 2008; C. A. *58,* 9114 (1963).
687) Wöhler, F.: Liebigs Ann. Chem. *127,* 257 (1863).
688) Wolf, E., Herbst, C.: Z. Anorg. Allgem. Chem. *347,* 113 (1966).
689) Wolf, E., Schönherr, M.: Z. Chem. *2,* 154 (1962).
690) Wolf, E., Teichmann, R.: Z. Anorg. Allgem. Chem. *347,* 145 (1966).
691) Wu, T. C., Gilman, H.: J. Org. Chem. *23,* 913 (1958).
692) Yamamoto, K., Kumada, M., Nakajima, J., Maeda, K., Imahi, N.: J. Organometal. Chem. *13,* 329 (1968).
693) Yamamoto, K., Okinoshima, H., Kumada, M.: J. Organometal. Chem. *23,* C 7 (1970).
694) Yamamoto, K. Okinoshima, H., Kumada, M.: J. Organometal. Chem. *27,* C 31 (1971).
695) Yamazaki, K., Kotera, A., Tatematsu, A., Iwasaki, M.: J. Chem. Soc. Japan *69,* 104 (1948).

696) Zintl, E., Bräuning, W., Grube, H. L., Krings, W., Morawetz, W.: Z. Anorg. Allgem. Chem. *245,* 1 (1946).

Received September 1973, literature till 1972

GPSR Compliance

The European Union's (EU) General Product Safety Regulation (GPSR) is a set of rules that requires consumer products to be safe and our obligations to ensure this.

If you have any concerns about our products, you can contact us on ProductSafety@springernature.com

In case Publisher is established outside the EU, the EU authorized representative is:

Springer Nature Customer Service Center GmbH
Europaplatz 3
69115 Heidelberg, Germany

Batch number: 10620163

Printed by Printforce, the Netherlands